Reference
Not For Circulation

Publication of this book is made possible, in part, through the generous support of the following:

The Fred A. Groves Endowment, Southeast Missouri State University

The Grants and Research Funding Committee, Southeast Missouri State University

and

The McCasland Foundation, Duncan, Oklahoma

Water Mills of the Missouri Ozarks

Water Mills of the Missouri Ozarks

By George G. Suggs, Jr.

With Paintings and Illustrations
by Jake K. Wells

University of Oklahoma Press : Norman and London

Library of Congress Cataloging-in-Publication Data

Suggs, George G., 1929–
    Water mills of the Missouri Ozarks / by George G. Suggs, Jr. : with paintings and illustrations by Jake K. Wells. — 1st ed.
      p. cm.
    Includes bibliographical references.
    ISBN 0–8061–2259–5 (alk. paper)
    1. Water mills—Ozark Mountains Region—History.   2. Water mills—Missouri—History.   3. Ozark Mountains Region—Industries—History. 4. Missouri —Industries—History. I. Wells, Jake. II. Title.
TJ859.S89   1990
621.2′1′097788—dc20   89–25082

The paper in this book meets the guidelines for permanence and durability of the Committee on Production Guidelines for Book Longevity of the Council on Library Resources, Inc. ∞

Copyright © 1990 by the University of Oklahoma Press, Norman, Publishing Division of the University. All rights reserved. Manufactured in the U.S.A. First edition.

To the memory of Owen John George Suggs and Jeanette Wells

* Contents *

Preface	xiii
Introduction: The Legacy	3
Water Mill Vignettes	39
Alley Spring (Old Red) Mill	43
Bollinger (Burfordville) Mill	47
Byrnesville Mill	53
Cedar Hill Mill	59
Dawt Mill	63
Dillard Mill	69
Dolle (Bollinger-Dolle) Mill	74
Drynob Mill	80
Falling Spring Mill	86
Greer Mill	90
Hammond Mill	94
Hodgson (Aid-Hodgson) Mill	99

Jolly (Isbell) Mill	104
McDowell Mill	108
Noser Mill	112
Old Appleton (McLain) Mill	116
Ritchey Mill	123
Schlicht Mill	128
Topaz Mill	132
Zanoni Mill	137
Epilogue: Going to Mill	143
Notes	165
Glossary	175
Selective Bibliography	181
Index	195

✳ *Illustrations* ✳

Plates

1.	Alley Spring (Old Red) Mill, Shannon County	42
2.	Bollinger (Burfordville) Mill, Cape Girardeau County	48
3.	Byrnesville Mill, Jefferson County	55
4.	Cedar Hill Mill, Jefferson County	58
5A.	Dawt Mill, Ozark County	65
5B.	Dawt Mill, Ozark County	67
6A.	Dillard Mill, Crawford County	71
6B.	Dillard Mill, Crawford County	73
7A.	Dolle (Bollinger-Dolle) Mill, Bollinger County	75
7B.	Dolle (Bollinger-Dolle) Mill, Bollinger County	77
8A.	Drynob Mill, Laclede County	81
8B.	Drynob Mill, Laclede County	83
9.	Falling Spring Mill, Oregon County	87

10. Greer Mill, Oregon County	91
11A. Hammond Mill, Ozark County	95
11B. Hammond Mill, Ozark County	97
12. Hodgson (Aid-Hodgson) Mill, Ozark County	100
13. Jolly (Isbell) Mill, Newton County	105
14. McDowell Mill, Barry County	109
15. Noser Mill, Franklin County	113
16A. Old Appleton (McLain) Mill, Perry County	117
16B. Old Appleton (McLain) Mill, Perry County	121
17. Ritchey Mill, Newton County	124
18. Schlicht Mill, Pulaski County	129
19. Topaz Mill, Douglas County	133
20A. Zanoni Mill, Ozark County	139
20B. Zanoni Mill, Ozark County	141

Figures

1. Millstone	5
2. Down by the Old Mill Stream	7
3. The Mill	9
4. Overshot Waterwheel	11

5. Greek Mill	17
6. Norse Mill	20
7. Cog gearing in a Roman Mill	23
8. Roman Mill	25
9. Tub Mill	33
10. Underwater Turbine	36
11. Going to Mill	151
12. Fishing While Waiting Turn	154
13. Receiving Mail at the Mill	157
14. Playing Horseshoes at the Mill	160

Map

Locations of Wells's Water Mills	41

Preface

It is not often that a historian has an opportunity to collaborate with an artist as talented as Professor Jake K. Wells, whose watercolors and sketches adorn this volume. Thus in the late seventies when the opportunity to join with him in producing this work presented itself, I enthusiastically seized it. Professor Wells, a first-rate regional artist who taught many years at Southeast Missouri State University, thereafter graciously and patiently accepted long delays in my preparation of the manuscript because of a longstanding commitment to another study then underway.

Water Mills of the Missouri Ozarks originated during the summer of 1977 when Professor Wells painted twenty-seven watercolors of twenty Missouri water mills. (The Jake K. Wells Collection is now owned and exhibited by the University Museum, Southeast Missouri State University.) All of these old structures were located in fourteen widely scattered counties of the Ozarks that extend in a sweeping arc from Franklin and Jefferson counties near Saint Louis southward to Cape Girardeau and Bollinger counties and then westward across the state to Barry and Newton counties (see map, p. 41). With a very few exceptions, the subject water mills were located in the most beautiful and most isolated regions of the state. The extraordinary natural environment of the millsites, of course, greatly enhanced the visual poetry that is found in the Wells paintings.

When selecting his subjects, Professor Wells did not envision a volume such as this one. His choices, therefore, were based on his artistic appreciation for particular structures, his lifetime fascination with specific water mills, his desire for artistic diversity, his fondness for the old rather than the new, and his wish to preserve an attractive part of Missouri's heritage. Thus the paintings included in this volume not only reflect a bit of Missouri's past but also reveal much about Wells and his artistic values. They disclose, for example, a gifted, sensitive man concerned with the slippage into obscurity of these ancient structures that he values so dearly.

During the summer and fall of 1986, Professor Wells and I visited the water mills that he had painted nine years earlier. Sad to say, several of these structures are no longer standing. They have fallen victim to indifference, neglect, and "progress." Nevertheless, in our imagination we recaptured the mills as they were, felt them throbbing with life, and visualized all the sights and sounds associated with their operations. We also renewed our appreciation of rural Missouri and its people. We were refreshed anew by all the natural wonders of Missouri's Ozarks.

In preparing this book, my principal purpose was to accent the artistic contrubutions of Professor Wells with historical vignettes—not exhaustive histories—of his mills. Therefore, following a brief introduction to place the mills into a historical framework, I have provided for each a short yet reliable account of its physical location, its origin, its significant owners, its singular characteristics, its modifications in time, its role in local affairs, and its present status. I have concluded with general observations about the water mills and their place in the lives of the people they served. This book is therefore not a history of Missouri's milling industry. Nor is it an explanation of the technical aspects of water-mill construction and operation over the years. The nature of the work precluded that kind of historical effort, and besides, it would be repetitious. Because of the difference in the evolution and history of the water mills that Professor Wells elected to paint, more information was available about some of his choices than about others. This accounts for the uneven treatment of the mills in the vignettes.

Finally, it should be noted that in the early decades of this century there were scores of water mills scattered about Missouri. This book briefly discusses only twenty of this vast number. It points to the great need for a thorough and comprehensive history of the early mills and their dynamic role in the state's growth and development. Such a history is long overdue.

In preparing this volume, Professor Wells and I have incurred many obligations. The following people graciously granted interviews, provided materials, assisted, and generally encouraged the completion of the work: Allen Mortimer, Bert and Margaret Wells, Mike and Nancy Cracraft, Mary Helen Flentge, Joe O'Neal, Art and René Dellamano, R. C. ("Rip") Schnurbusch, John and Jeredie Nesbit, Tay Smith, Jack Smoot, Dorothy Krieger, Erwin Viehland, Edna Lewis, Duard Johnson, Linda Johnson, Ellen Gray Massey, Kirk Pearce, Tony Czech, Gorton Thomas, Manford Troxel, Emory Styron, L. Orville Goodman, Curmel ("Red") James, and James Lalumondiere. We are grateful to them for their encouragement and help.

Provost Leslie Cochran, Vice-presidents Robert Foster and A. R. Meyer, Deans David Payne and Sheila Caskey, Chairpersons B. G. Ramsey and Charles Bonwell, and Museum Director James V. Parker of Southeast Missouri State University strongly supported our work from its inception. We gratefully acknowledge their assistance. We are grateful to the university and its museum for permitting the Jake K. Wells Collection to be reproduced in this book, and to Paul Lueders, master photographer, who prepared superb color transparencies from the paintings. Our thanks go also to the staff of Kent Library, Southeast Missouri State University, whose members, especially Betty Black and Maribeth Needels, were most generous with their time and talent. Professor Wells and I are further indebted to the university for his paid leave during the summer of 1977 and for my sabbatical leave during the fall semester of 1986 which made this book possible. We owe a special debt to Janice Hart for her patience and help on the computer. We acknowledge with appreciation the help and support of family members who know full well how much they contributed to this book.

GEORGE G. SUGGS, JR.

Cape Girardeau, Missouri

*Water Mills of the
Missouri Ozarks*

* Introduction
The Legacy *

By the start of the twentieth century, the water mill in the United States was well in decline as a primary source of power, its glory days having gradually faded with the spread of roads and highways and the introduction of new technologies such as the steam engine, which powered more and more of America's factories and mills. Nevertheless, for generations of rural Americans growing up in the early decades of this century, the water-powered mills continued to exercise a profound influence on their lives despite a diminished importance in doing the nation's work.

The continuing influence of water mills on the American mind was less economic than artistic and aesthetic, a fact noted by O. B. Bunce long before the dawn of the twentieth century. Writing in 1872 about the nearly forgotten mills that he had found scattered along the Brandywine in hidden and inaccessible places, in *Picturesque America* Bunce vividly described the gripping fascination that the old water mills exercised on the imagination of artists and poets: "What is there in an old mill by a brook that fascinates so quickly the eyes of an artist and the heart of a poet? . . . Probably no object in Nature or art has been so often drawn and painted. And yet, familiar as we are with old mills nestling quaintly among summer foliage, we always discover a fresh fascination in each new example. Was there ever an artist who could resist the desire to add a new sketch of a [water mill] to his portfolio?"[1]

Bunce was undoubtedly responding as much to the natural settings of the old mills (for example, the brook and the summer foliage) as to the mills themselves. Nevertheless, he sensed the haunting beauty and serenity resulting from that remarkable combination of ingenious craftsmanship, architectural form, and the natural elements that remained the enduring attraction of old mills to the early generations of the new century.

But if the era of the water mills left a legacy of obsolete and crumbling structures possessing the power to stir the hearts and the imagination of poets and artists, like the mills included in this volume, it also left a dying linguistic residue, a small lexicon of colorful words and phrases that has enriched the everyday language of the modern masses. Even before the heyday of the water mills in the western world, these expressive verbal tidbits had already crept into significant literary

Fig. 1. Millstone

works, such as the *Bible*, that tremendously influenced the language of later generations.

To illustrate his teachings, Jesus drew freely upon his natural and cultural surroundings, which included hand-, slave-, cattle-, and water-powered mills that had been developing long before his birth. On one occasion he warned that it would be better for anyone to have a millstone (Fig. 1) "hanged about his neck" and to drown in the depth of the sea than to offend an humble believer.[2] Although the millstone was probably not from a water mill like that illustrated, the vividness of that Biblical injunction combined with the growing prevalence of water mills and the religiosity of the next fifteen centuries to incorporate into the language the phrase "millstone around the neck"—meaning, of course, any weighty or seemingly insurmountable problem. In our contemporary world, who has not experienced a problem so "heavy" that it seemed like a "millstone around the neck?" And is there anyone who does not grasp the meaning of the phrase when it is heard?

Another residue of the mill era became embedded in the nation's musical past. In 1910, composer Tell Taylor published "Down by the Old Mill Stream," a popular tune which continues to be a favorite among barbershop quartets whose members, more than likely, have never seen a water mill. Like Bunce nearly a half-century before, Taylor sensed

Fig. 2. Down by the Old Mill Stream

the powerfully romantic aura created by the old water mills and their environs. For Taylor they resonated with poetry and song. As Wells's sketches (Figs. 2, 3) suggest, in such an idyllic and romantic setting it was natural that a young man and a gingham-dressed, blue-eyed, sixteen-year-old village queen should meet and fall in love. It was also natural that he should rhapsodize in verse and song about the place where his heart awakened to love. For Taylor's contemporaries, "Down by the Old Mill Stream" was rich with sentiment and meaning, its imagery conjuring up familiar places. Today, it generates little response because, except for a few restored showpieces, old water mills nestled by running brooks and streams are nearly extinct.

But for Jake Wells, whose sketches and watercolors adorn this volume, the water mills were more than legend and song. Very early in his youth, he fell under their spell, a spell that has persisted for a lifetime. A much admired grandfather was a miller who worked in the water mills near Marble Hill in the Ozarks of southeastern Missouri where Wells grew up. As a child, Wells visited his grandfather at work, played around the mills, and became captivated by the water and the natural elements surrounding his grandfather's workplace.

With his father, Wells later visited mills on the upper reaches of Missouri's Black River, the southeastern boundary of the Missouri Ozarks.

Fig. 3. The Mill

On one of these visits he experienced for the first time the sights and sounds of a flume-fed, overshot waterwheel much like the wheel of the Zanoni Mill, paintings of which are included in this volume (Plates 20A, B). He was intrigued by what he saw and heard. "To watch that wheel turn," he later recalled, "just fascinated the dickens out of me. Now, I don't know whether the thing actually was operational or whether it was just some sort of cosmetic type of thing, but after I had seen this mill, I began looking around for any opportunity I had to go to other water mills. There were quite a few of them back then." There followed trips to the Bollinger (Plate 2), Baker [McManus], and Dolle (Plates 7A, B) mills, which were located, according to Wells, "all within a country block" of Marble Hill. Thereafter, Wells was never able to escape the powerful, aesthetic pull of the water mills, even those without the external wheel like the one sketched for this volume (Fig. 4).

In time, the mills became a natural artistic subject for Wells's brush because they were compatible with his development as an artist. "I have a feel," explained Wells, "for old buildings that seem to be on the way down as opposed to something that's spic-and-span and well kept." The accuracy of that assessment is evident in his watercolors of Missouri mills, for time has ravaged most of the mills into an advanced state of

Fig. 4. Overshot Waterwheel

ruin. His paintings of the Drynob (Plates 8A, B) and the Hammond (Plates 11A, B) mills beautifully reveal his responsiveness to the old. His "feel" for these skeletal, bleak reminders of a bygone era clearly permeates his work. When viewing Wells's paintings, one finds confirmation of Bunce's observation in 1872 that "whether the mill be one quaint and fantastic by virtue of its decay and ruin, or one that lifts its walls from the river-edge in large pretension, there is always a strange pleasure in this combination of the beautiful and the useful."[3] Wells's paintings of Missouri's old water mills generate in the viewer the same pleasurable response noted by Bunce. (For example, see his paintings of the Dawt Mill [Plates 5A, B] and the Schlicht Mill [Plate 18].) That they do so is a tribute to Wells's talent and sensitivity to the beauty lurking in these old structures and their settings.

If Wells's paintings convey his feeling for time-worn architectural forms, perhaps it is because within the painter there lurks a poet. In visiting the mills around Marble Hill as a youth, Wells perceived in each more than an inanimate structure composed of mortar, wood, and stone. Instead, he sensed that each was a living thing possessed of soul and spirit. In reminiscing about the Bollinger (Plate 2) and other mills, he recalled that while in operation the mills throbbed with life—that to the sensitive visitor they spoke an unmistakable language of form, function, spirit, and personality. In his view, water mills were living things endowed with uniqueness and individuality.

"One thing that did interest me—or fascinate me—about the Bollinger mill, as well as others," he said, "was the motion once you were inside the mill. Regardless of what floor you were on, if the mill was in operation you could feel the beat or vibration throughout, with the elevators going all through the mill. I don't know whether you could compare it to being on a ship, but in its way, it had life, or a kind of rhythm to it." Mills, like living things, had much in common in terms of form and function, but each mill, he believed, also had its own personality—something that made it unique. The poet in Wells found expression in his mill paintings.

In harnessing the energy of Missouri's plentiful rivers and streams, the builders of the state's water mills left a legacy of crumbling structures that appeal especially to artists and other lovers of beauty. Wells's sensitivity to these architectural relics and to their rustic sites makes his paintings unusually appealing. However, the water mill, with its origins in ancient times, was not developed to satisfy the demands of man's aesthetic nature. It was instead a response to necessity rather than the needs of poets and artists. This was certainly true for the mills of the Missouri Ozarks that Wells has painted.

The water mills of Missouri, which began their decline toward ruin in the late nineteenth century, did not spring from a vacuum. Nor did they originate in the inventive genius of a frontier people struggling to tame a stubborn continental wilderness. On the contrary, the conception of the mill technology used to open up Missouri (especially its Ozark region) and the rest of North America occurred in the dimness of ancient times. For centuries it had been evolving from the murky shadows of a transatlantic past, with generation after generation of scattered peoples adding their own unique refinements.

After the discovery of America in the late fifteenth century, for more than three hundred years the water mill, like other cultural and mechanical forms, rode the crest of European migration to the continents of the Western Hemisphere. Of all the elements of western civilization borne by early immigrants, perhaps none had greater practical significance than the tested and proven water mill. It was an ingenious mechanical legacy that made the task of taming the sprawling continent a little easier for waves of sturdy pioneers.

Yet historians, after probing the origins of the legacy, remain unable to identify with certainty the inventors of the water mill and to determine when and where it first appeared. Their investigations, which have included a careful combing of ancient Western texts for references to its use, produced only tidbits of information indicating a history of slightly more than two thousand years—a mere moment in the long history of mankind. But the conception of pounding, crushing, or rolling cereal grains into a more edible form—an essential idea in developing the water mill—extended more deeply into the past.

Perhaps as much as ten thousand years ago people began to domesticate cereal grains only to find that their achievement was a mixed blessing. Although more conveniently located and harvested, the more abundant domesticated grains had a hard outer shell that quickly wore away the teeth. Necessity therefore dictated a preliminary preparation to make them more edible. And probably from observing the grinding and crushing action of human and animal jaws and teeth, man derived the idea of using stone or wood as pounders or crushers. Primitive pounding devices (for example, the pestle and mortar) were the result.

From these early hand "mills" evolved more sophisticated ones (all essential precursors to the water mill itself), such as the revolving quern and the slave- and cattle-powered mills. These simple mills embodied the basic idea of using moving stones to grind corn and other cereal grains into a more edible form. Each employed a fixed bottom stone (bed stone) and a revolving upper stone (runner stone) powered either by humans or animals. Conceiving and applying the idea of harnessing the power of running water rather than that of humans or animals to turn millstones represented a tremendous leap forward in mill technology. Some historians believe that leap had occurred about 85 B.C., or approximately a century before the birth of Christ. It set the stage for more extensive developments.[4]

Two types of ancient water mills then existed. One was the Greek, an inefficient horizontal contrivance that was extremely easy to build. The other was the Roman, a more efficient mill that was comparatively complicated and difficult to construct. Among practical, rural people, the former's lack of sophistication (for example, the absence of cog gearing (see Fig. 7) made it readily adaptable to their limited demands. Its simplicity is evident from the following description:

> The water-wheel [of the Greek mill] lay [parallel] upon or in the water, and revolved an upright central shaft standing upon a stone in the bed of a stream, or else in a dry channel to which water was conveyed by a trough. The upper end of the shaft or spindle passed through the lower of the two quern-like grinding-stones placed above, but was fixed to the upper stone. Thus the water-wheel, the shaft, and the upper stone all revolved together [Fig. 5]. Such a mill, erected upon a small stream, would only grind very slowly, as one revolution of the water-wheel would of course only produce one revolution of the grinding-stone.[5]

Fig. 5. Greek Mill

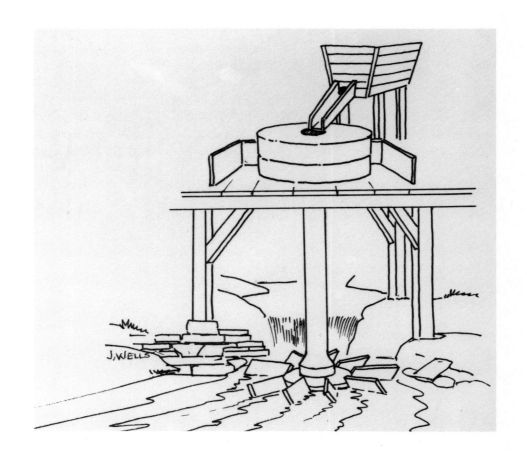

Adequate, cheap, and easily built, the Greek mill spread both east and west. For nearly eight hundred years it was widely used throughout Europe until it was generally displaced by the more efficient Roman type sometime between the eighth and the twelfth centuries.[6]

A variation of the Greek mill, however, extended much further into time. From the seventh to the sixteenth centuries, an era known as the Middle Ages, the Norse mill flourished in northern and western Europe. Found in such widely dispersed places as Ireland, Scotland and Norway, it was so similar to the Greek mill, and so equally primitive, that a structural description of the one portrayed the other and suggests a common origin for the two. For example, compare the description of the Greek mill above with the following accounts of the Norse mill, the first written in the fifteenth and the other in the seventeenth century: "In the Isle of Man, on many of the rivers, is a cheap sort of mill, which, as it costs very little, is no great loss though it stand idle six months in the year. The water-wheel lies horizontal, consisting of a great many hollow ladles, against which the water brought down by a trough strikes, and gives motion to the upper stone, which, by a beam of iron [shaft], is joined to the centre of the water-wheel."[7]

It is evident that the general structure of this mill was the same as that of the Greek type. The same was true of the following mill situated in Ireland: "In county Down issue many rills and streams, and on almost each of them a townland had a little miln [mill] for grinding oats. . . . [T]he axel-tree [shaft] stood upright, and [the] small stones (or querns, such as are turned with hands) on [the] top thereof. The water-wheel was fixed at [the] lower end of [the] axel-tree, and did run horizontally among [the] water, a small force driving it [Fig. 6]."[8]

Although environmental circumstances often imposed infinite variations (for example, the dimensions of the waterwheel, the number and size of the ladles or paddles inserted into the wheel, the length and substance of the shaft, and the measurements of the millstones), the basic form of the Greek and Norse mills nevertheless remained the same. It consisted of four parts: a paddled or ladled horizontal waterwheel, a vertical shaft, a fixed lower stone, and a revolving upper stone, or runner stone, attached to the shaft. And when properly combined, they constituted a very primitive but functional water mill which ingeniously harnessed a bit of nature's power to relieve human drudgery. Such mills proliferated well into the age of exploration beginning in the fifteenth century and, in slightly modified form, came to North America as part of the cultural baggage of European immigration.

The second of the water mills known to exist around 85 B.C. was the Roman. In their conquest of the ancient world, the Romans created a

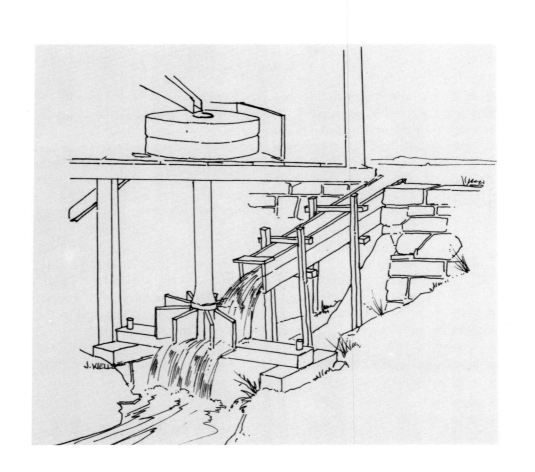

sprawling empire that encompassed much of the Mediterranean basin. Their conquests exposed them to the scientific and cultural achievements of the peoples they subjugated. Among the conquered, of course, were the Greeks with their horizontal water mill. The Romans, a practical-minded people, perceived the hidden potential in the Greek mill and made important innovations which increased its efficiency. However, because the Romans long retained a definite preference for slave- and cattle-powered mills, their creative innovations were not widely used—even within the empire itself—until centuries later, when circumstances forced them to change.[9]

Following Christianity's entrance into Rome and the resulting reduction of slavery and slave-powered mills, the growing demands of the populace for meal toward the end of the fourth century compelled the Romans to explore the productive potential of the more efficient water mill that they had all but ignored for several hundred years. Its increasing use and importance were soon reflected in a growing body of laws that regulated millstreams and protected the mills.[10] Consequently, by the eighth century the Roman mill had gradually begun to supersede the more primitive Greek or Norse mill throughout northern and western Europe. And from it came the prototype of the old water mill with its external wheel.

Fig. 6. Norse Mill

Except for added innovations, the constituent parts of the Roman mill were the same as those of its Greek predecessor. There was a waterwheel, but instead of being positioned horizontally, it was now set upright. There was a connecting shaft, but it was placed horizontally rather than vertically. Vital to each mill, of course, were the two millstones—the lower, fixed bed stone and the upper, revolving runner stone. Their positions in the Roman mill remained unchanged from that in the Greek.

The Roman innovation that substantially advanced mill technology was the introduction of cog gearing (Fig. 7) to alter the direction of mechanical power through the mill. It was this development which combined with the upright waterwheel and the horizontal shaft to plumb the extraordinary hidden potential of the water mill. To the horizontal shaft the Romans fixed a cogwheel (face wheel) with sturdy cogs or teeth inserted at intervals into the rim of its interior face. These cogs intermeshed at right angles with a lantern pinion, sometimes called a wallower (a gear with bars inserted between the rims of two wheels to form a cylinder), that was fixed to a vertical shaft connected to the upper millstone. Consequently, as water turned the wheel, power flowed through the horizontal shaft, the cogwheel, the lantern pinion, and the vertical shaft to run the upper millstone.[11]

Fig. 7. Cog gearing in a Roman Mill

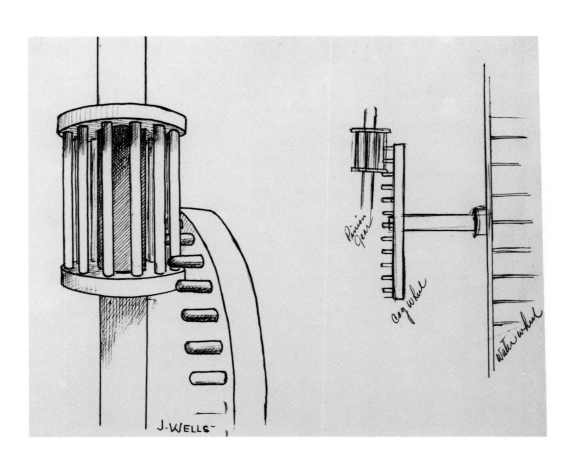

The remarkable thing about incorporating cog gearing into the power train of the mill was that by altering the number and/or size of the teeth in the cogwheel or the lantern pinion, the runner stone could be made to revolve faster than the waterwheel itself. Thus, the turning speed of the stone was no longer limited to that of the waterwheel. Cog gearing greatly complicated the flow of power through the vital parts, but it also greatly enhanced the productive capacity of the water mill. More sophisticated both in form and function, the Roman mill (Fig. 8) possessed the potential for applications far beyond the grinding of grain.[12] Later generations fully exploited that potential.

Rome's expansion eastward into Greece had confronted it with an advanced culture whose achievements it selectively appropriated and modified for its own use. An example of that acquisition process was the water mill. But in its conquests to the north and west in Europe, Rome bore a culture superior to that of the peoples it subjugated. Throughout the conquered provinces as far west as Britain, its roving legions left a cultural imprint that lingered on long after they had withdrawn. There, language, architecture, law, religion, the practical arts—all were influenced in one way or another by Roman conquest and occupation. Nevertheless, historians and others continue to dispute the extent of that influence.

Fig. 8. Roman Mill

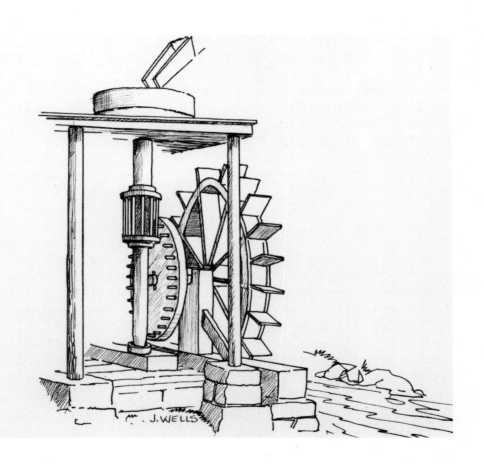

One such disagreement concerned the water mill: When did it first appear across the English Channel in faraway Britain, and was it really a Roman import? Over the years speculation has ranged from an outright rejection that it was anything but an indigenous development to the candid admission that no one knows for certain its origins in Britain. In their classic *History of Corn Milling*, Richard Bennett and John Elton concluded that the "Romans did not introduce the Greek mill" into Britain because they had not "troubled to establish their own better and more powerful [Roman] mill" there during their stay as conquerors. Why? Even in Rome itself, they note, no serious development of waterpower occurred until after 398 because of a preference for slave- and cattle-powered mills. Furthermore, they speculated that Britain's first awareness of the water mill was probably in the fifth century, when Teutonic tribes swept over the British Isles, likely leaving the Norse mill in their wake. If the Romans introduced their mill into Britain, according to Bennett and Elton, they likely did so in the fifty years between 398 and 448, when they withdrew their legions to Rome. However, little physical or legal evidence (mills, millstones, or mill statutes) has been found to support the conclusion that they did so.[13]

One finds a somewhat different view in Terry S. Reynolds's more recent *Stronger Than a Hundred Men*. Reynolds concluded that for "Britain there is no evidence of watermills between the collapse of the Roman Empire and the eighth century, when a document of King Ethelbert of Kent (762) mentions a watermill." By the ninth century, however, the water mill was "known over wide areas of England" and it was used in Wales by the tenth century. Although Reynolds did not speculate about the appearance of the Roman mill in Britain, he concluded that during the "period of approximately 600 years between the invention of the water wheel and the complete collapse of the Roman Empire in the West (c100 B.C. and c500 A.D.) the use of water power had apparently not spread beyond a few sharply restricted areas," which did not include Britain.[14] Nevertheless, by whatever means, the Roman water mill eventually found its way into Britain and over time gradually supplanted the widely used Norse mill.

By the ninth century, water mills had become a highly visible, practical institution in many communities of northern and western Europe. On the banks of the Voire in France, the monastery of Montieren-Der built eleven water mills to serve the inhabitants of its twenty-three domains. And the monastery of Saint Germain-des-Pres constructed fifty-nine water mills on the streams that meandered through its lands. During the course of the next several centuries, the number of water mills in France and elsewhere tremendously increased. In one department (Aube), according to one account, there were only fourteen mills in the eleventh century, but sixty in the twelfth and two hundred or more in the thirteenth. In Paris during the fourteenth century, sixty-eight mills were situated within a mile of each other along the Seine River.[15]

Across the English Channel in Britain during the six centuries following the withdrawal of the Romans, the water mill, increasingly the Roman type, also flourished until by the end of the eleventh century its use was common throughout the land. In 1085, William the Conqueror, the Norman who had overrun Anglo-Saxon England in 1066, heard rumors that his new domain was about to be invaded from Scandinavia. Needing to know the financial resources available to meet this threat from the north, he ordered in 1086 a comprehensive survey of his kingdom. When completed, it provided a remarkable statistical inventory of the holdings of land, plows, mills, meadows, freemen, slaves, and so on, found in nearly every locale. Furthermore, it revealed substantial information about the industries and resources that existed there even before William's conquest in 1066. One extraordinary revelation of the survey, later compiled and known as the *Domesday Book*, was the existence of numerous mills scattered throughout England both before and during the reign of the Conqueror (1066–1087). Although "completely unrecorded" until 1086, the remarkable proliferation of water mills used for grinding grain during the four hundred years before William has been described as one of that era's "greatest economic achievements."[16]

Of the 13,418 settlements included in the *Domesday* inventory, slightly more than 25 percent (3,550) had one or more mills. Altogether, a total of 6,082 mills were uncovered by the survey. Approximately 25 mills were found for every one thousand households. Because there is no documentary record of a windmill in England before 1191, the *Domesday* mills were most likely water mills of the types described above, with the Roman type probably predominating.[17]

During the next several centuries, developments in England and the rest of Europe encouraged the use of the larger and more efficient Roman mill. The need to centralize the administration of large estates, the feudal practice of claiming exclusive jurisdiction over milling (that is, the practice of making peasants use the landowner's mill), and the growth of towns caused the Roman to supplant the less efficient Greek or Norse mills by the twelfth century. Of equal importance was the increasing use of water mills for tasks other than grinding grain. Before the end of the fourteenth century, a more generalized application of waterpower had occurred in textiles (fulling mills), dye making (crushing mills), ore processing (stamping mills), and lumbering (sawmills). According to Reynolds, by 1500 "medieval European society [had] enthusiastically adopted water power and incorporated it into the predominant feudal-manorial system. If the ancient world gave birth to the vertical water wheel and nurtured the earliest stages of its growth, it was the medieval West that brought it through adolescence and into adulthood."[18]

Yet, that "adulthood" could not have been attained without other supportive developments that made the Roman water mill the principal part of an expanding power system. Mention has already been made, for example, of cog gearing, which greatly advanced mill technology. In addition, at the end of the fifteenth century the knowledge of how to construct and use dams, reservoirs, millraces, sluice gates, flumes, chutes, and tailraces to drive the mills was generally known throughout Europe. Such knowledge made possible the creation of what Reynolds called "hydropower complexes" and a "very substantial growth in industrial dependence on water power during the Middle Ages."[19]

Thus, before the exploration and settlement of North America had begun, a water-powered machine of enormous potential had evolved among the Europeans, especially the English, who were to use it to master the North American continent in the centuries ahead. Long before Missouri had been defined, the technological basis for many of the the water mills that today nestle here and there in the countryside of its Ozarks had substantially evolved.

Even so, in his *Waterpower: A History of Industrial Power in the United States, 1780–1930*, Louis C. Hunter concluded that while colonial records from before American independence reveal much about the "introduction, increase, and spread of water mills," they reveal little about "construction and equipment." For example, there is virtually no information about the kinds of waterwheels employed, how the wheels were built, and how they were linked to mill machinery. With two exceptions during the colonial era, Hunter wrote, "We do not know what variations upon Old World forms and arrangements may have been introduced, whence and by whom, what modifications in the details of design, construction, and practice may have been promoted by colonial conditions and needs." Those exceptions were the Hammersmith Ironworks of Saugus, Massachusetts, and a gristmill built in a Swedish community in eastern Pennsylvania. Both facilities were constructed in the 1640s.[20]

Among the first large-scale corporate enterprises in British North America, the water-powered Saugus works were a rather sophisticated, integrated iron mill consisting of a "blast furnace, a bloomery forge, and a rolling and slitting mill, the last for making nailrods." Vertical overshot wheels were used to drive the machinery. Although this venture eventually failed, one historian has suggested that its technology was equal or superior to the best to be found in Europe.[21]

From the standpoint of early Missouri water mills, however, the gristmill of the Swedes was more significant. Investigations of the millsite suggest that it was a tub mill (Fig. 9), a variation of the Norse mill that was widely used in Sweden. If that is true, Hunter speculates, this "may well mark the coming of this important type to the United States."[22] Because of its simplicity, compactness, and low cost, the tub mill was probably the type first used when settlers penetrated wilderness areas like the Missouri Ozarks.

Both in structure and operation, the tub mill was essentially a horizontal mill of the Norse type (described above) with a primary difference. Its wheel, which was nearly identical to that of the horizontal mill, revolved in a tublike wooden enclosure without a bottom. According to Charles Howell, "From a flume or sluiceway, a downwardly inclined, enclosed wooden trough or penstock was tapered inwardly toward the delivery point and directed the water in a jet against the blades of the wheel. The protruding sides of the tub formed a continuous apron to

Fig. 9. Tub Mill

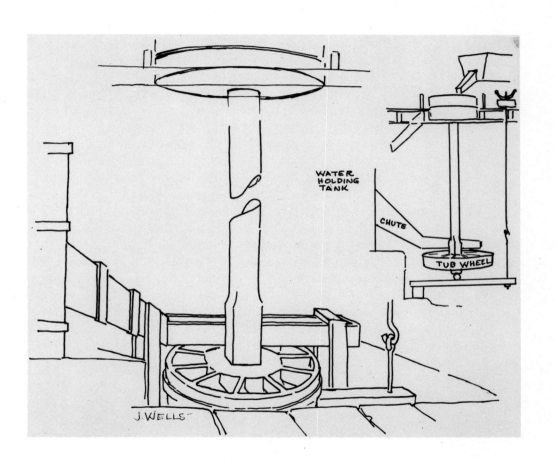

prevent the water from escaping sideways, which improved the efficiency of the wheel."[23] The tub mill was a forerunner of the turbine.

Together with simpler horizontal hand- and animal-powered mills, the tub mill was generally used in backwoods America (as such mills had been used in early Europe) because of its simple construction and ease of operation. In time, however, many of these mills were replaced by the more efficient vertical waterwheel mills of some kind. Of course, the latter were replaced in turn by the more efficient underwater turbine. (Steam, electricity, and diesel engines later replaced the turbine.) As the following vignettes of Ozark mills will show, Missouri millers converted from the external wheel to the turbine whenever possible.

Introduced into the United States in the 1840s, the turbine by 1880 had generally displaced all forms of the external wheel. It was a product of much experimentation, principally in Europe, and combined the characteristics of the tub wheel and the reaction wheel. (The reaction wheel was one that operated submerged, either in a horizontal or vertical position. Water rushing through the wheel struck its passages or buckets and generated reactive pressure or force to drive the wheel.) Innovators blended into the turbine the "simplicity, compactness, and high rotating speeds" of the tub wheel with the "ability [of the reaction wheel] to run submerged and in either a horizontal or vertical position." The result was a wheel of far greater efficiency than all the others discussed above [Fig. 10].[24]

Hunter described the turbine that reached the United States in the early 1840s as follows:

> The new turbine was a relatively simple mechanism with three principal components: a central fixed disk on which were mounted a number of iron guides that curved downward and outward, forming spiral passages by which the water passed from the penstock to the wheel proper; a horizontal wheel, or runner, mounted on a vertical shaft and having two outer rims, separated by vertical metal strips dividing the space between them into a number of curved passsages, or buckets, through which the water received from the fixed guides moved outward; and a gate mechanism by which the admission of water from the penstock to the wheel was regulated. . . . [T]he turbine's buckets . . . presented curved surfaces against which the water exerted force by pressure and reaction in passing through the wheel. . . . In the turbine all the working surfaces were simultaneously subject to the pressure of the column of water passing through the wheel. Like the small horizontal wheel preceding and culminating in it, the turbine was fundamentally a reaction wheel, its power being derived principally from the reactive pressure of the water upon the surfaces of the buckets or passages from which it issued.[25]

Most of the highly automated roller mills that are described and illustrated in the following vignettes (for example, the multistoried Topaz [Plate 19] and Bollinger [Plate 2] mills) were driven by similar turbines installed during the post–Civil War era.

To a large extent, such automated mills were made possible by the work of Oliver Evans, a young American inventor of the late eighteenth and early nineteenth centuries who brought his genius to bear on the problem of grinding grain. Following years of experimenting, he constructed outside Philadelphia a water-powered mill in which, through the compact integration of conveyors, elevators, and so on, production was completely automated. Using the Evans process, grain was introduced into the system, moved by conveyors and chutes through the various phases of milling and refining, and emerged as a finished product of flour or meal. Except to start it up, Evans's mill required little labor, a fact that tremendously reduced the cost of production and hurried its acceptance.

In 1795, Evans published his *Young Mill-Wright and Millers' Guide*, which became the standard manual for entrepreneurs like Captain Samuel W. Greer, who built the Greer Mill (Plate 10). By 1850, Evans's book, described as the "Miller's Bible" because it was so widely used, had gone through thirteen editions.[26] Consequently, much of the milling

Fig. 10. Underwater Turbine

technology brought into the Missouri Ozarks in the post–Civil War era was derived from that book.

When Evans's technology was combined with other innovations of the nineteenth century, such as turbines and steel rollers, which replaced the millstones, the efficiency of the mills and the quality of their products were further enhanced. Often these later improvements were the result of external conditions that necessitated change. For example, the development and switchover to rollers resulted from newer strains of hard wheat, introduced in the postwar era, that buhrstones could not adequately process. Such improvements revolutionized milling and made possible large-scale production for distant markets. As the following vignettes will show, Ozark millers were quick to accept and install these innovations.

Settlers moving westward into the Missouri Ozarks therefore carried with them the milling technology of centuries developed abroad *and* in North America. It helped them survive in an isolated, wilderness environment which, fortunately, possessed the natural characteristics favorable to the water mill.

Water Mill Vignettes

The Ozarks region is a sprawling, sometimes rugged land of approximately sixty thousand square miles found in Missouri, Arkansas, Oklahoma, Kansas, and—according to some sources—Illinois. Its boundaries are roughly defined by major rivers: on the north, the Missouri; on the east, the Mississippi; on the southeast, the Black; on the southwest, the Arkansas; and on the west, the Grand and the Neosho. The northwestern boundary, running from just north of Joplin, Missouri, northeastward to Howard County, Missouri, is defined by an ancient rock formation. In size, the region is larger than Arkansas, with more than half of its land area situated in Missouri, where Wells's mills were located. In addition to distinguishing geological features (for example, an abundance of dolomite and chert), the Ozarks country is physically characterized by a rough, hilly topography; generally average to poor soils; fast-flowing streams; forests of oak, hickory, and pine; and many caves, springs, and sinkholes.[1]

All the conditions, including an annual rainfall of from thirty to fifty inches, were present to produce in the Ozarks the "basic elements of waterpower—volume of flow and amount of fall." As Louis Hunter noted about many places in North America, in the Ozarks there was "no part of the land . . . without its running water, no stream however small without its power potential. For running water is falling water and falling water has only to be supplied with dams, races, and waterwheels, in some instances by waterwheels alone, to serve the needs of industry and man."[2] Consequently, when the French and Spanish began to penetrate the Ozarks with their watermill technology in the seventeenth and eighteenth centuries, they found abundant sites for their mills. And as successive waves of settlers from the Carolinas, Kentucky, and Tennessee pushed deeper into the region during the nineteenth century after the United States acquired ownership, watermills, although often very primitive in construction, proliferated to serve their milling needs. An estimated 850 mills in 1870 and 900 in 1880 existed throughout Missouri. Of course, not all of these were water mills,[3] but among them were simple country mills, such as Falling Spring Mill (Plate 9) and the highly complex, automated merchant mills such as the Bollinger (Plate 2) and Topaz (Plate 19) described in the following vignettes.

Locations of Wells's Water Mills

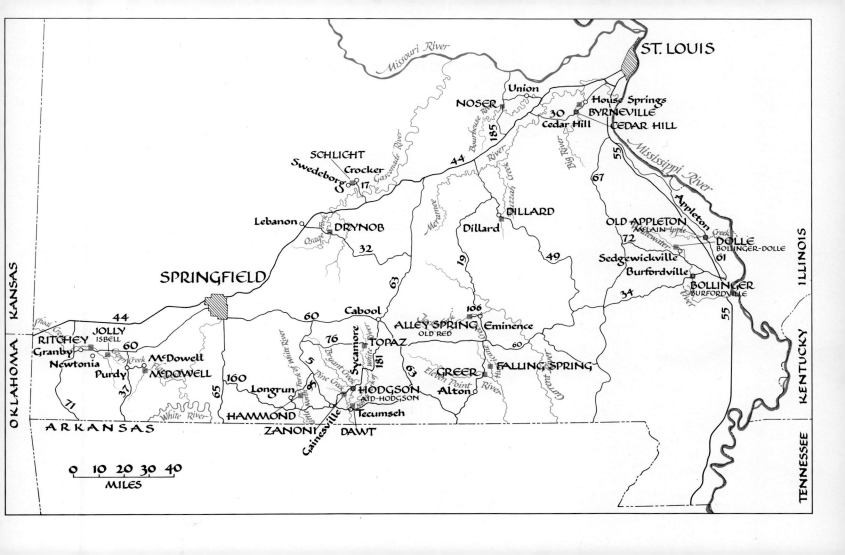

✱ *Alley Spring (Old Red) Mill* ✱

Located on State Highway 106 six miles west of Eminence in Shannon County, the Alley Spring Mill (often called the Old Red Mill) stands near an extraordinary natural phenomenon—a natural spring whose average daily flow exceeds eighty million gallons (see map, p. 41). Surfacing at the base of a limestone bluff, the spring forms a deep pool, the overflow of which moves on to join the Jack's Fork River. At the spring in 1870, Ike Barksdale, a blacksmith, and his partner, John Dougherty, constructed a dam and built a nearby gristmill with millstones (also called buhrstones) powered by a wooden vertical waterwheel that Barksdale also made. Like so many other early mills in isolated areas of the Ozarks, this gristmill was a custom mill that was limited in capacity and ground only the small grain production of local farmers for a toll in kind—that is, for a percentage of the product. It complemented a country store and blacksmith shop belonging to the partners. (A blacksmith

Plate 1. Alley Spring (Old Red) Mill, Shannon County

shop was usually found near a water mill like the Alley Spring. It was essential to repair or to refashion damaged parts quickly and on site.) Thus, the milling, smithing, and mercantile needs of the local inhabitants were met by Barksdale and Dougherty, who operated these properties for twelve years.

Charles Klepsig, a Prussian immigrant with several sons who owned a farm on nearby Jack's Fork River, acquired the interests of the partners in 1881. Unfortunately, Klepsig's family did not maintain the properties, especially the mill, and allowed them to run down. They were in a deteriorated state when acquired by George McCaskill in 1893.

McCaskill poured his resources into rebuilding the mill into a merchant mill, that is, one with a capacity to produce for consumers outside the immediate area. He repaired the milldam, installed an underwater turbine, an encased, water-powered rotary engine with redirecting vanes and passages that allowed the intake and release of water as desired, to replace the old vertical waterwheel, and replaced an old shed that had covered the millstones with the present structure, which is portrayed so beautifully in Wells's painting. McCaskill also supplemented the millstones with modern steel rollers to make a better quality of flour. By adding the turbine and the rollers, McCaskill increased the potential production of the mill to twenty-four barrels of flour and fifty bushels of meal per day. At the same time, he continued to operate the store and the blacksmith shop that were acquired with the property.

Unfortunately, McCaskill's innovation created an unneeded milling capacity. To operate its new rollers efficiently, the mill needed large quantities of grain to produce flour for sale beyond nearby markets. Local farmers, however, rarely produced grain in excess of family needs. Consequently, in 1912 an investor from Kansas City acquired the mill properties, hoping to convert them into a desirable resort for urbanites. He failed. By 1918, the mill ceased to operate.

During the ownership of Barksdale and Dougherty, the gushing spring which powered the mill had come to be called Barksdale Spring. The name changed, however, when McCaskill requested John Alley, a farmer on the Jack's Fork River, to move the post office that he operated to the settlement around the spring. Alley agreed to do so. Locating the postal service in a gristmill or in a nearby general store was not uncommon. Because mills were frequently the focal point of community life, the government often appointed their owners postmasters. Sooner or later most area residents found it necessary to go to the mill to replenish their stock of meal or flour. Being able to receive their mail while there was a great and an appreciated convenience. The Alley Spring Mill, like its counterparts throughout the Missouri Ozarks, was an agent of community. While waiting their turn at the mill, farmers who lived isolated from each other used the time to socialize. They caught up on recent events in the lives of their neighbors, learned of happenings elsewhere, and reestablished contacts with distant family members and friends who might be at the mill. The water mills therefore made it possible for local residents to do their milling, horseshoeing, mailing, and buying all at the same place.

In time, local inhabitants called the spring Alley Spring. The formal name change, however, did not occur until 1948, when, as a result of a successful petition requesting the change, the spring officially became Alley Spring. Part of the Alley Spring Mill, which is now administered by the National Park Service, is operative for demonstration purposes. The remainder is a museum.[4]

✳ Bollinger (Burfordville) Mill ✳

The history of the Bollinger Mill extends back into the Spanish and French period of the Mississippi River valley, making it one of the oldest water mills in Missouri. In 1797, George Frederick Bollinger, a former North Carolinian who found the Whitewater River country to his liking, received a land grant of 640 acres from Don Louis Lorimier, the Spanish commandant who administered the Cape Girardeau district. Lorimier granted this section of land to Bollinger on the condition that Bollinger would develop it and populate the surrounding area with other immigrants.

In 1799, before returning to North Carolina to recruit settlers, Bollinger constructed a dam across the Whitewater River and a log gristmill on limestone pillars at a point later to be called Burfordville. These structures were primitive compared to those of stone, brick, and mortar now on the site. Nevertheless, they aided tremendously in persuading pioneer families to move to Missouri. Returning to North Carolina in

1800, Bollinger recruited twenty families to move west and settle along the Whitewater River near his property. The new residents, of course, greatly enhanced the value of the mill, and they launched more than a century and a half of nearly uninterrrupted milling on the site in Bollinger County.

The Bollinger Mill, like so many early Missouri water mills, evolved rather slowly into its present form. The original mill probably was powered by an external waterwheel for nearly a quarter of a century. In 1825, Bollinger, then well-to-do and involved in state politics, established his milling business on a more permanent basis. He rebuilt the mill, replacing the supporting pillars with a solid, walled-in foundation of limestone blocks. He used wood for the upper stories and limestone blocks for a new dam. After his death in 1842, his properties, including the mill, were placed in trust for his daughter Sarah until 1850, when she acquired full control following the death of her husband, Frank Daugherty, whom Bollinger greatly disliked. The second mill continued to operate until Union forces torched it during the Civil War.

In March, 1866, Sarah Bollinger Daugherty sold the remains of the mill and other property to Solomon R. Burford for $12,400. Burford, after whom the town of Burfordville is named, built a third mill of stone and brick. This is the present structure, the rear of which, together with

Plate 2. Bollinger (Burfordville) Mill, Cape Girardeau County

the dam and covered bridge, Wells painted from a vantage point across the Whitewater River. It was powered by a forty-inch external wheel until 1879, when, upon Burford's establishment of a partnership with L. Silverman to form S. R. Burford and Company for the purpose of producing flour, a more powerful underwater turbine was installed as a condition of the partnership. Other improvements, such as the addition of rollers, were probably made in order to diversify and enhance the quality of products. In the fifteen years between 1880 and 1895, Burford's heirs disposed of their interest in the mill, fractional shares being acquired by various individuals. Consequently, milling there ceased to be solely a family affair.

Reflecting the consolidating tendencies of the times, in 1895 the Bollinger Mill and two flour mills in nearby Jackson, Missouri, were consolidated to form the Cape County Milling Company. Fully automated, the mill continued to operate until 1953. It produced a high-quality bleached flour using a revolutionary process developed in 1903 by James N. Alsop of Jackson. Although merged into a larger corporate entity whose markets far exceeded anything dreamed of by Bollinger, the Bollinger Mill retained its distinctive hold on the region's imagination.

Ownership in the historic old mill continued to change after production ceased. In 1953, the family of Paul Vandivort, descendants of

George Frederick Bollinger, purchased the property from the Cape County Milling Company. Eight years later the Vandivort family transferred ownership to the Cape County Historical Society. Unable to restore the mill because of inadequate resources, the society transferred title in 1966 to Cape Girardeau County. Next year the Bollinger Mill and the magnificent covered bridge partially shown in Wells' painting were accepted by the Missouri Department of Natural Resources. The Bollinger Mill State Historic Site, with the mill as the centerpiece, was the result.

Located in Burfordville about fifteen miles west of Jackson off State Highway 34 in Cape Girardeau County, the Bollinger Mill remains a remarkable structure nestled on the Whitewater River (see map, p. 41). Its history demonstrates the remarkable transformation that water mills in Missouri sometimes underwent over time. That is, it originated as a small "custom" gristmill that ground corn meal on order for a toll in kind and culminated as a large merchant mill with a capacity to produce flour and meal for distant markets. Like so many other of the state's water mills, it promoted the origin and growth of a town that came to include mill-related and other needed businesses: a blacksmith shop, a sawmill, a general store, a post office, and so on. Fortunately, because of the new status of the mill as a state historic site, the town lives on.

Furthermore, the existence of the mill undoubtedly made the Whitewater River region more attractive to early pioneer families, thereby stimulating its growth. Over the years the mill has been subjected to the natural hazards of all water mills—fire and flood—and the man-made hazard of war. In spite of it all, the four-story Bollinger Mill, as Wells's watercolor indicates, remains today a beautiful and stately edifice that graces the Whitewater River countryside.[5]

* Byrnesville Mill *

The history of the Byrnesville Mill began in 1847 when David Manchester applied for a state permit to construct a dam across the Big River in Jefferson County in order to provide power for a water mill that he planned to build. State permission was required to assure the continued commercial use of the river and to determine whether the dam would result in flooding of nearby farmland. The permit was granted. Consequently, when Manchester built his milldam, which was nearly twice as long as any other then on the Big River, he provided a rafting chute to accommodate the timber cutters who moved their product down the river. When the Missouri Ozarks region became a major source of railroad ties after the Civil War, such chutes were necessary where the rivers were used for moving them. Constructed of timber cribbing, a box-like structure of logs usually filled and backed by stones, this original dam has endured remarkably well, having been breached and repaired only once since its construction.

Following the building of the dam, Manchester constructed a mill about 1850. It was most likely made of logs. Its power plant was probably an external waterwheel of some type. Between 1847 and 1867 when Patrick Byrne acquired the mill, it went through a series of owners, for a while changing hands about every two years. When Byrne purchased the mill with its 350 acres, it was known as Yerkes Mill. At that time Byrnesville consisted of the mill and a general store.

Upon acquiring the property, Byrne completely rebuilt the mill. As portrayed in Wells's painting, it was a three-story frame structure that rested on a walled-in stone foundation and projected straight up from the Big River. (In the post–Civil War era, many of the new water mills were built with three or four stories in order to fully integrate an automated milling process.) In addition to the millstones, Byrne added rollers to meet the new demands of an increasing population in the region. It appears that even with these renovations, the mill continued to operate from a waterwheel, possibly a low-efficiency undershot wheel, which the water struck at a point below the axle.

Patrick Byrne did not live long after buying the mill. He died in 1872, leaving the property to his son, M. F. Byrne. Afterwards, the Byrnesville Mill was owned and operated by different members of the Byrne family until 1903. While the mill was still owned by the Byrnes, another major

Plate 3. Byrnesville Mill, Jefferson County

improvement was added to it in 1882. An underwater turbine, purchased from a company in Springfield, Ohio, was installed to provide greater efficiency in milling operations. A custom mill, the Byrnesville Mill principally served the needs of the local population, its owners generally receiving a toll in kind for the services rendered. Nevertheless, there is reason to believe that sometime during its history, its flour, under the brand name Lilly White, reached markets as far away as Festus.

Like so many Missouri water mills, the Byrnesville Mill gave rise to a small community. By 1875 the village of Byrnesville consisted of the mill, the store, a blacksmith shop, four dwellings, and a post office. M. F. Byrne was the first postmaster. During the next fifty years, other small businesses were added (for example, additional blacksmith shops) to serve a growing population in the area. Having originated with the mill, the town depended upon its continued operation for business. For nearly a century the mill attracted large numbers of farmers from the surrounding area and sustained the economic life of the community and local residents.

Between 1903, when Charles Klienschmidt purchased the mill, and 1936, when its operations totally ceased, its ownership changed several more times as the need for small gristmills and flour mills gradually diminished. Like similar mills throughout Missouri, it fell victim to external changes beyond the control of local millers. Thus, when milling operations ended in 1936, the stoppage was permanent. For forty years, from 1936 until 1976, the Byrnesville Mill lay silent and rotting, interrupted only by curiosity seekers and by the partial gutting of its machinery for scrap iron during World War II. Its decay was accompanied by the decline of Byrnesville as a thriving community.

Located southwest of House Springs off State Highway 30 on Byrnesville Road, the Byrnesville Mill has now taken on new life (see map, p. 41). During the past eight years, James Lalumondiere, who acquired ownership in 1976, has gradually transformed the old mill into a substantial residence. Although the structure has been radically altered, Lalumondiere has preserved the essence of the mill, both inside and outside. Through his efforts its useful life has been extended.[6]

✻ *Cedar Hill Mill* ✻

During the nineteenth century, a number of water mills were built on the Big River, with three of them situated on that portion of the river flowing through Jefferson County. Most of them, like the Byrne Mill (not to be confused with the Byrnesville Mill), located about five miles northwest of House Springs, have all but disappeared. A deteriorating milldam, a stone and concrete foundation with pillars, and a half-submerged underwater turbine with its vertical shaft intact are all that remain of the Byrne Mill, a once thriving enterprise that was vital to the people it served. Although only partially successful in its relentless assault on the remnants of the Byrne Mill, nature nevertheless has successfully reclaimed other millsites along the Big River. An important exception is the Cedar Hill Mill, located in Cedar Hill on State Highway 30 about thirty-five miles southwest of Saint Louis (see map, p. 41). The history of this mill, which remains operational, reaches back nearly

Plate 4. Cedar Hill Mill, Jefferson County

150 years. For nearly 100 of those years, the Radeacker family of Cedar Hill retained possession of the property.

County records indicate that the first mill on a nearby site originated in 1847 when Thomas Maddox petitioned the county court for the right to construct a dam across the Big River. His intent was to build a gristmill and sawmill and use the water behind the dam to power them. His petition provoked immediate opposition from Conrad Beehler, who owned property across the river. Beehler was concerned that the dam would cause flooding and damage his interests. Nevertheless, when a jury appointed by a state circuit court ruled against Beehler, Maddox proceeded to construct a log and stone dam and to build the mills. These are thought to be the first water mills ever built on the Big River in Jefferson County. Whether Maddox used a waterwheel or an underwater turbine to drive the machinery is uncertain. Although milling continued on the site for more than thirty years, not much is known about its operations until ownership of the millsite passed into the hands of Louis Radeacker. There is some suggestion that the old mill (for a long time known as the Maddox Mill) fell victim to either fire or flood, for it was rebuilt in 1876.

When viewed from across the Big River, as in Wells's painting, the Cedar Hill Mill is a stately, three-story frame building resting on a stonewall foundation that rises from the river near the site of the Maddox Mill. Built by Radeacker about 1890, the mill was powered by underwater turbines that at one time turned both imported French buhrstones (prized millstones made from siliceous stone) and rollers positioned in the upper floors. The turbines continued to power the mill until 1954, when the owners began using a diesel engine for supplementary power. Later, for greater reliability there was a total shift from water and diesel power to electricity, a move which ended the mill's reliance upon energy derived from the Big River. It is no longer a water-powered mill.

Over the years production gradually shifted from a concentration on meal and flour to mixed feeds for hogs, cattle, and chickens, but changes in the feed business eventually forced the owners to switch production substantially to pet foods. By the early 1980s, pet foods made up approximately 50 percent of the products sold. During the 1940s, the mill also began producing block ice, which was in heavy demand to preserve food in old iceboxes. The sale of ice became a lucrative business. Its production, however, accelerated the changeover from water power to the more reliable diesel engine and electricity. Ice continues to be produced, but less so now than in the 1950s and '60s, when demand was great.

Upon building the Cedar Hill Mill, Louis Radeacker operated it with the help of his seven sons. Upon his death, the mill became the property of sons William and Albert. In 1937, Wilbert Radeacker purchased into a partnership with William, and in 1950, William sold out his interest to Walter Radeacker. Wilber and Walter Radeacker continued to operate the mill (now known as the Cedar Hill Ice and Feed Mill) until 1974, when they sold the business, but not their mill property, to Erwin Viehland, who had worked in the mill for many years. In 1984 the business changed hands again when Viehland retired. James Lalumondiere, who has plans for using the mill to generate electricity, purchased the property in 1982.

Although the Cedar Hill Mill ceased to be a water mill upon conversion to diesel power and then electricity, its interior reflects its earlier reliance on waterpower. The underwater turbines are still in place, and shafts and pulleys from the old days remain available for use should it again become necessary to draw energy from the river. Occupying a commanding, panoramic view of the Big River, after nearly a century this working mill continues to provide a useful service to residents of Cedar Hill and the surrounding area.[7] And as Wells's watercolor suggests, it also adds an aesthetic dimension to the surrounding landscape.

* Dawt Mill *

It would be difficult to find a more suitable and beautiful setting for a water mill than the site of the Dawt Mill a few miles north of Tecumseh in Ozark County (see map, p. 41). To exploit the enormous waterpower potential of the magnificent North Fork of the White River at the upper end of Norfork Lake, Alva Hodgson, a builder who had previously constructed the Hodgson Mill and other water mills, erected the three-story Dawt Mill in 1900 on a slight bend on the east bank. He then constructed a sturdy, hand-built dam of stone and mortar across the wide river, angling it upstream to redirect the swiftly flowing water toward the millrace and the turbines. Angling the dam, as was often done with mills on substantial streams (for example, the Old Appleton Mill on Apple Creek, the Noser Mill on the Bourbeuse River, and the Cedar Hill Mill on the Big River), added stability and strength against floods and to resist ramming by floating trees and logs and increased the power of the water. Viewed from the west bank of the North Fork, the dam and the mill are conspicuous structures that nevertheless blend harmoniously into the river environment. Wells has captured the essence of that harmony in his painting of the back of the mill from the water's edge.

Like that of so many of Missouri's existing water mills, the history of the Dawt Mill site before 1900 is shadowy and vague. For example, there is even uncertainty about how the mill got its name. Furthermore, one source suggests that milling occurred on the location as early as 1867 on land first privately owned by Rhuhama J. Isom, who acquired a patent to the acreage from President Ulysses S. Grant in 1874. But, according to another source, John Caldwell built and operated the original gristmill there as late as 1897 in competition with the Friend Mill, which was situated several miles upstream. A story prevails that the owner of the Friend Mill, intolerant of the downstream competition, decided upon drastic measures to retain his monopoly of the local milling business. Taking advantage of a "simple-minded youth," the owner paid him ten dollars to burn out his downstream competitor in 1900. However, when the lame mule which he had ridden to the mill was tracked to his home, the youth confessed the arson and exposed the owner of the Friend Mill as the instigator. It is reported that Alva Hodgson immediately began constructing the present mill on the old site, using parts of the original

Plate 5A. Dawt Mill, Ozark County

foundation. There is some evidence that the Dawt Mill is located on the site of the earlier mill.

Upon completion of the mill, Hodgson, who with his family had lived near the Hodgson Mill at Sycamore for some time, moved to Dawt into a substantial farm house that he also had built. The house is still standing. For a while, Hodgson operated the mill, which, like many other Missouri mills, developed a number of auxilliary businesses. These included, of course, a blacksmith shop, a general store that supplied basic household items and dry goods, a cotton gin, a post office located in the store, and, with the coming of the automobile, a filling station. In short, the little town of Dawt developed around the mill.

Except for the times when the mill ceased to operate because of flooding, a break in the dam which necessitated repairs, or water erosion of the foundation that required attention, the Dawt Mill has continued to operate since 1900. Powered by underwater turbines and using millstones and rollers, it was the area's largest producer of meal and flour.

Although the Dawt Mill continues to fill small custom orders for meal, it has changed greatly in function and appearance since it was built. A nonfunctional overshot waterwheel has been added for the tourists, with water being pumped from the river to turn it. (Wells's frontal painting of the mill, however, does not show this wheel and other recent changes

Plate 5B. Dawt Mill, Ozark County

made to attract tourists.) And Phil Horne and Tay Smith, present owners, have outfitted the underwater turbines to produce small quantities of electricity. The old auxilliary businesses of cotton ginning and blacksmithing have been replaced with camping and canoeing outlets. Although it has changed, the Dawt Mill on the North Fork River remains a thing of beauty.[8]

Dillard Mill

In 1853, eight years before the outbreak of the Civil War, Francis Wisdom built a water mill on Huzzah Creek near the present-day towns of Dillard and Viburnum in Missouri's Central Ozarks. Powered by a waterwheel and known as the Wisdom Mill, it stood for more than half a century until partially burned in 1895. Before the fire, one of its several owners was Joseph Dillard Cottrell, after whom the town of Dillard was named. Today it is known as the Dillard Mill.

In 1900, Emil Mischke and his sister Mary, Polish immigrants who had settled in the Dillard area during the 1890s, acquired the mill property. Four years later they began constructing a new water mill, which was completed in 1908. The Mischke Mill was very different from the original Wisdom Mill. A substantial structure, it employed the newer technologies, including an underwater turbine, steel rollers, and a purifier. (Lennis L. Broadfoot, a self-trained artist who painted the mill in the late 1930s or early '40s, included a picturesque waterwheel where the underwater turbines are now located. Furthermore, Broadfoot wrote about the mill's "giant overshot water wheel." No such wheel existed.) These innovations required extensive alterations in the millsite. The Mischkes reworked the bluff on which the mill had been built to make possible a new log dam, a new millrace to conduct water to the turbines, and a rerouting of Huzzah Creek. The altered bluff can be seen in Wells' painting of the rear of the mill. In 1917, Emil Mischke became the sole owner when he bought out the interest of his sister. His ownership continued until 1930, when Lester Klemme, a businessman from Webster Groves, Missouri, purchased the property.

Changing demands in milling during the Great Depression forced Klemme to switch the primary production of the mill from flour to cattle feed. To generate much-needed additional income, he also capitalized on the natural beauty of his property and the surrounding Ozarks by establishing the Old Mill Lodge, where paying guests resided in small log cabins, fished and swam, and ate home-cooked meals with the Klemmes. Below the milldam was a deep, bowllike basin filled with crystal clear water, so skillfully captured in Wells's watercolor. It was the site of much social activity, for there the local residents enjoyed picnics and basket dinners, swimming, and fishing. Enhancing the beauty of the millsite were trees of all kinds: sycamores, maple, walnut, black and

Plate 6A. Dillard Mill, Crawford County

white oak. As Wells's frontal view of the mill reveals, these trees continue to flourish there.

Throughout its history the Dillard Mill was called by several names—Wisdom Mill, Mischke Mill, Klemme's Dillard Roller Mill, and finally Dillard Mill, acquiring the last name when the state assumed control in 1975. Though the mill ceased operations in 1956, the Klemmes exercised a strong, protective watchcare over their property. They kept its equipment intact, maintained the building, and warded off potential vandals. Klemme transferred ownership of the mill to the Leo A. Drey Foundation in 1974, and the next year it became part of the Missouri Department of Natural Resources, Division of Parks, Recreation, and Historic Preservation, through a lease arrangement with the foundation. It was dedicated in 1977 as the Dillard Mill State Historic Site. The restoration of the mill began in 1979 and was completed in the following year. The Dillard Mill is located off State Highway 49 near Dillard and Viburnum in Crawford County (see map, p. 41).[9]

Plate 6B. Dillard Mill, Crawford County

Dolle (Bollinger-Dolle) Mill

Mathias Bollinger was the brother of George Frederick Bollinger, who built the original Bollinger Mill at Burfordville and with it enticed settlers into the Whitewater River country. Among those George lured to Missouri was Mathias, also a Protestant from North Carolina, who early in the 1800s constructed a gristmill using water from the Whitewater River to power its waterwheel and imported French buhrstones.

The Dolle Mill, like the hilltop Greer Mill in Oregon County described later, was located a considerable distance from the river that fed its waterwheel. To reduce the danger of flooding, a constant danger to most watermills, Bollinger selected a slightly elevated millsite about two or three miles from the river and off its flood plain. To create the essential head of water to operate the machinery of the distant mill, Bollinger constructed a cribbed log and stone dam across the Whitewater, thereby raising its level so that water could be diverted into a millrace for a slightly downward run.

Plate 7A. Dolle (Bollinger-Dolle) Mill, Bollinger County

For a millrace, Bollinger dug a canal about eight feet wide and four feet deep from the river to a millsite pond, from which water was directed to the wheel. The foreground of Wells's rear-view painting of the mill suggests the volume of water carried by the diversion canal. Bollinger reportedly used slaves brought to Missouri from North Carolina to provide the tremendous amount of manual labor required to construct the millrace and dam. Nevertheless, in transmitting waterpower from the distant river to the mill, he displayed notable engineering skill and ingenuity. Furthermore, in doing so he made the Dolle Mill one of the most unusual among Missouri water mills.

Sometime between the building of the mill and 1853, Mathias Bollinger and his son Moses, who had fallen heir to the property, died. In order to divide equally the estate of Moses among his heirs, in 1853 the mill property was sold at a sheriff's sale on the steps of the Bollinger County courthouse at Marble Hill. The buyer was John Herman Dolle, who had emigrated with his family from Prussia to the United States in 1837. After acquiring ownership, five succeeding generations of Dolles owned and operated the works. They retained possession until 1937, when the Bollinger family again regained possession of the mill. Ever since it has remained a Bollinger property.

Like the Bollinger and other mills, the Dolle Mill was threatened by

Plate 7B. Dolle (Bollinger-Dolle) Mill, Bollinger County

the movement and clash of opposing military forces during the Civil War. Dolle, who probably fled Prussia to escape military service, nevertheless enlisted in the Union Army. While he was away in service, his wife Mary ran the mill. When Confederate forces operating in the region seized the property and learned that its owner was a Union soldier, they prepared to burn the mill. When the officer in charge, a Mason, noticed a Masonic emblem worn by Mrs. Dolle, he countermanded the order to burn. In this manner, according to family history, the Dolle Mill was saved from deliberate destruction.

Located four miles northwest of Sedgewickville on County Road EE in Bollinger County (see map, p. 41), the Dolle Mill was one of several that served the people in its immediate area. Situated on a winding, unpaved road with its back to the millrace dug by the slaves of Mathias Bollinger, the impressive two-story frame structure remains in the sound condition reflected in Wells's paintings. There is some suggestion that the present mill is not the original but a replacement that was built following a legendary fire caused, as the story goes, by a pigeon carrying a burning substance into the rafters. Sometime in its early history, an underwater turbine replaced the old waterwheel.

Like so many of Missouri's mills, the Dolle Mill was a social center for the people it served. For a while it was the Masonic hall and post office.

Historically a custom mill that ground wheat and corn into flour and meal on demand with payment in kind, the Dolle Mill never evolved into a substantial milling operation like the nearby Bollinger Mill, which supplied faraway markets. Nevertheless, it has endured as an operating mill with very limited functions. No longer a water mill, it is powered by a gasoline engine and produces primarily mixed feeds.[10]

✱ Drynob Mill ✱

A half-dozen or more water mills were built here and there in Laclede County during the nineteenth century, but perhaps the most interesting was the Drynob Mill located on Osage Fork River about thirteen miles east of Lebanon off State Highway 32 (see map, p. 41). Before its construction, a more primitive gristmill had been built below the present millsite. This original water mill was replaced by the Drynob Mill and dam.

Though the mill is thought to have been constructed in 1890, its builder and early owners have become obscured by the passage of time. One source points to Allen Parham as the builder and operator before the turn of the century. Parham is thought to have sold the mill property to John McElroy, whom one old-timer remembers as operating the works during her youth. McElroy was credited with rebuilding the original into an up-to-date flour mill whose product was much superior to that of its predecessor. Despite these uncertainties, the three-story

Plate 8A. Drynob Mill, Laclede County

frame structure was well built, for it withstood the assaults of time until October, 1986, when it collapsed backward toward the creek, a victim of old age and neglect. Wells's paintings reveal the complex timbering in the foundation that sustained the mill for nearly a century.

During the Civil War, many of Missouri's mills with waterwheels were destroyed by contending forces as they swept through the state. When these mills were rebuilt, except in the most isolated places, nearly all of them incorporated the latest power technologies, using underwater turbines or steam to drive the machinery rather than the primitive waterwheel. When the Drynob Mill was built, however, its builder resorted to the wheel to turn its buhrstones. He probably did so because, unlike the turbine, which was usually produced elsewhere, it was cheaper and could be built on-site. Edna Lewis, an old-time resident of the town of Drynob, remembered the wheel because of its size and the sight and sound of its splashing water. It continued to be used to power the mill until 1946, when it was replaced by more efficient underwater turbines.

At that time the mill was owned by Douglas Stout, who had recently purchased it from Bill Starnes. Stout apparently intended to expand and improve milling operations, for with the installation of the turbines he also installed rollers to improve the quality of the flour produced at the works. A carpenter, Stout also worked on the mill as if he intended to

Plate 8B. Drynob Mill, Laclede County

run it on a permanent basis, adding a new roof and making other repairs. Shortly after finishing these renovations, however, he stopped milling operations in 1949 and moved to Oklahoma. The following year, the family of long-time resident Linda Johnson acquired the property, including the land, the building, and the machinery that remained intact inside the mill. The Johnsons retained ownership of the Drynob Mill until its collapse in October, 1986, which, unfortunately, buried all the machinery in the debris. Rollers, shafts, and other equipment were left protruding haphazardly from the wreckage.

From the time of its construction until milling ceased in 1949, the mill was one of the centers around which the community of Drynob developed. Farmers for miles around took their corn and wheat there to be ground into meal and flour. Linda Johnson recalled that these products were of excellent quality. As a consequence of the farmers going to mill at Drynob, other businesses also developed there to serve them.

Two general stores, a blacksmith shop, a post office, and several houses composed the town near the mill and dam. While waiting their turn at the mill, farmers fished, visited the blacksmith for harness repair or horseshoe replacement, picked up their mail, or shopped in stores that were stocked with essential dry goods and basic commodities. Thus, going to mill provided widely separated inhabitants of the region an opportunity to renew acquaintances and transact necessary business without having to go to distant Lebanon by mule and wagon. When external developments, including improved highways and roads, forced the closing of the Drynob Mill in 1949, the community that it helped to spawn began its slow decline. Like the mill which gave it birth, the town of Drynob is now extinct.[11]

Falling Spring Mill

Early water mills were often located in isolated places, their location being determined solely by the source of water. Some, however, were more isolated than others, their existence perhaps known in the early days only to the people in the area they served. The Falling Spring Mill appears to have been such a mill. It is located deep in the southwest corner of the Mark Twain National Forest in Oregon County. The mill can be found by traveling north from Alton over State Highway 19 until reaching an unmarked graveled road seven miles beyond the Eleven Point River bridge. By traveling east on this road into the forest and bearing left when it forks, one can locate the mill about four miles off Highway 19 (see map, p. 41). Within a half-mile of the millsite, the Falling Spring Cemetery, with gravestones dating back to the 1860s, attests to the community the mill once served.

In contrast to the automated, multistoried Dawt, Bollinger, Topaz, and similar mills, which were powered by underwater turbines, the Falling

Plate 9. Falling Spring Mill, Oregon County

Spring Mill was extremely small, with only one story, and operated from an overshot waterwheel. The water powering the wheel originated from a spring located in bluffs that rise some fifteen or twenty feet behind and above the mill. To use the spring, which overflowed through a high cleft in the bluffs, the builders constructed a long wooden flume to catch and redirect the water to the wheel. An inactive steel wheel about fifteen to twenty feet in diameter remains in place. It is a conspicuous feature in Wells's painting. (This wheel had been previously used at another mill called Johnson Spring Mill, located several miles away on Hurricane Creek.) The flume, however, is gradually rotting away, with the water returning to its natural flow.

Although the inside layout of most water mills did not provide for spaciousness, the miller at Falling Spring had even less room to maneuver. Exposed shafts and gears, which yet remain inside, occupied much of the interior. At best, it was a gristmill with limited productive capacity and was obviously designed to custom grind corn meal and feed for the livestock of local residents. A unique feature was the small millpond situated in front of the mill that served to receive rather than provide water to feed the wheel. The pond therefore was not involved in mill operations. It was most likely a fish pond.

There is reason to believe that another water mill rested on the site before the present one. The original mill was reportedly constructed about one hundred years ago, about the same time as an old log cabin located near the millsite. It was powered by a wooden waterwheel, and its auxilliary services, like the mill itself, were limited. It is very likely that at one time a small sawmill also operated off the original wheel.

J. W. Brown, whose parents had migrated into the region in 1853, acquired the five hundred acre tract on which the mill was located and constructed the present Falling Spring Mill during the late 1920s. Brown left its support beams to be enclosed later. His son Walter, later heir to the property, also operated the mill. During the 1940s he used the mill to power a sawmill, to recharge batteries, and to perform other farm-related tasks. In 1957, Walter sold the mill property to Emil Slovak. Several years later, it was incorporated into the Mark Twain National Forest.

The Falling Spring Mill was probably typical of the water mills that proliferated throughout the isolated Ozarks in the early nineteenth century. Small of structure and limited in capacity, it nevertheless adequately served its clientele in the surrounding community. A picturesque relic in an idyllic setting, the mill has fallen victim, like its more sophisticated counterparts, to external developments that destroyed its reason for being.[12]

* Greer Mill *

Unlike most of Missouri's water mills, the Greer Mill is not situated directly near a body of water in a romantic and picturesque setting. Instead, it is located on a hilltop approximately three-fourths of a mile from and above the foliage-hidden Greer Spring, which furnished power when the mill was operational. Like the Dolle Mill, its distant location from its power source made it one of the most unusual water mills in the state.

Greer Spring, reported to be the second largest spring in Missouri, with a daily flow of over two hundred million gallons, was named after Captain Samuel W. Greer. Greer migrated from North Carolina to Tennessee in 1849 and, after a decade, moved on to Oregon County, Missouri, where he and his father, John, purchased land on which the spring is located. Samuel Greer, a carpenter and millwright, built the original gristmill next to the spring in 1859, shortly before the outbreak of the Civil War. Following military service for the South, he returned to find

Plate 10. Greer Mill, Oregon County

his mill destroyed by bushwackers. He rebuilt it, and about 1870 he enlarged it into a three-story structure, contructed a dam at the spring, installed an underwater turbine, and provided new facilities for the local residents: a sawmill, a cotton gin, and a carding mill.

In order to use these new improvements, however, heavily loaded wagons drawn by mules and oxen had to maneuver down a steep grade, a dangerous venture to say the least. And when the wagons were loaded with the finished products, teams and drivers confronted the equally hazardous task of moving them up the hillside. Consequently, when the present sturdy, three-story pine frame structure was begun in 1883, it was placed on the hilltop well above the spring, a most unusual location for a water mill but one that had safer access. Operation of the new mill began in 1899.

The distance from the spring necessitated an ingenious method of transmitting power from the turbine operating in the water below to the mill located at the top of the hill. With the help of George W. Mainprize, Greer developed a system that consisted of continuous steel cable strung on pulleys carried in three towers. The system extended down a swath cut through the forested hillside. Through this sophisticated drive system flowed the power generated by the gushing water of Greer Spring and its rushing branch that flowed on to join the Eleven Point

River less than two miles away. Remnants of the transmission cable can yet be seen near the rocky path leading to the spring.

   The present Greer Mill, completed around the turn of the century, continued to meet the milling needs of the local population until 1928, when production ceased. In the twentieth century the mill property has changed hands several times, being owned at one time or another by George Mainprize, Louis Houck, the Missouri Iron and Steel Company of Saint Louis, the Louis E. Dennig family, and Leo Drey. Although inoperative for more than fifty years, the now empty mill continues to attract tourists and others interested in Missouri's past. As Wells's painting reveals, Greer Mill is a majestic structure even in old age. It is located close to State Highway 19 approximately ten miles north of Alton in Oregon County on a magnificent tract of private property near the Mark Twain National Forest (see map, p. 41).[13]

Hammond Mill

The establishment of a water mill usually led to the emergence of a small village whose life depended upon the continuation of the mill. In order of appearance, the mill came first and was followed by a clustering of auxilliary businesses such as a blacksmith shop and a general store. The Hammond Mill was different in that it did not follow this customary pattern of development. S. J. Williams, John Squires, and John W. Grudier, founders of the little village of Hammond, selected the townsite because it was near an excellent location for a dam across the Little North Fork of the White River which had sufficient water to power a water mill. Thus, the potential of the site rather than an existing mill gave rise to the village. In time, this now extinct village, which mushroomed in 1907 and 1908, consisted of a whiskey distillery, a post office, two general stores, a drugstore, a bank, and, of course, the Hammond Mill.

Located on County Road D off State Highway 95 about four miles

Plate 11A. Hammond Mill, Ozark County

north of Longrun in the western part of Ozark County (see map, p. 41), the Hammond Mill at one time dominated the local countryside and economy. Built about 1910 by Grudier, a sawmill owner, the three-story mill of wood and stone was fully automated and was powered by underwater turbines that drove rollers and millstones. (A frontal view of the mill can be seen in Wells's painting.) Its products were flour, meal, and livestock feeds ground from the abundant wheat and corn produced in the area. Its flour sold under the brand name Ozark Queen Flour and was recognized locally as a superior product. Like its counterparts elsewhere, the Hammond Mill was busiest during harvest season, when many farmers arrived either on horseback or on wagons loaded with grain for milling. At such times they often found it necessary to wait for a very busy miller, who, when occasion demanded it, ran the mill day and night. In the meantime, the farmers socialized and perhaps fished in the millpond behind the mill.

Those busy times passed, erased by the growth of large, faraway milling companies with access to modern transportation and regional and national markets. The years since have been harsh on the Hammond Mill, which, when new, was a majestic building for the locale. After milling ceased there many years ago, little was done to maintain this last prominent structure of the village of Hammond, which has disappeared.

Plate 11B. Hammond Mill, Ozark County

Standing near an unpaved country road, the mill has gradually deteriorated. Although its stone foundation remains basically sound, its weatherboarding is falling off piece by piece, and wind and rain damage is slowly weakening its support beams. Underneath the mill, pulleys, shafts, gears, and support beams are scattered in disarray. Even the stream that powered its turbines has meandered off and deserted the mill, electing instead a different course and leaving the original stream bed nearly dry. Adding insult to injury, a strong tornadic wind in 1986 lifted the roof completely off and set it down neatly about fifty feet away. Nevertheless, the grand old Hammond Mill continues to defy time and the elements. Wells's painting of the embattled structure from a distant point captures that defiance. It stands tall and erect as if insisting upon survival against all odds.[14]

✱ *Hodgson (Aid-Hodgson) Mill* ✱

It is not often that an old water mill continues to have products marketed under its name in modern stores and supermarkets, but today packages of stone-ground white and yellow corn meal and wheat and rye flour bearing the Hodgson Mill brand can still be purchased. Although modern milling has all but eliminated the sale of such products, devotees of stone-ground meal and flour remain loyal to the Hodgson Mill and similar brands. And although Hodgson Mill products are now produced in nearby Gainesville (twenty miles to the southwest of the original millsite), each package of meal or flour with its picture of the old water mill is a reminder to the purchaser of a bygone era when such mills dotted the Ozarks in great numbers.

    Like the Alley, Greer, and Falling Spring mills, the Hodgson Mill relied upon a gushing spring to provide power for its operations. Snugly built against a steep bluff and positioned over the spring's outlet, the mill was strategically located to harness the enormous power generated by the

daily flow of nearly twenty-four million gallons of water. The original builder undoubtedly realized the potential of the site when he constructed a water mill there. After use, the spring water was discharged and channeled into a small, man-made millpond, situated in front of the mill, the overflow of which formed a branch that ran tumbling into Bryant Creek a short distance away. Like the millpond of the Falling Spring Mill (Plate 9), which was also frontally located, the pond of the Hodgson Mill had nothing to do with its operations. Wells has masterfully used the pond with its overflow to enhance the beauty of the mill in the background.

Located in the northeast section of Ozark County in the Sycamore community on State Highway 181 about twenty miles northeast of Gainesville (see map, p. 41), this great natural millsite deep in the Ozarks was originally occupied by a water mill built by William Holeman in 1861. Scant records reveal little about the history of the mill and its operations during the twenty-three years that it was owned by Holeman. There is, however, some evidence that the mill was forced to shut down during the Civil War. Nevertheless, the many years of Holeman's ownership and operations after the war suggest that he ran a profitable business.

In 1884, Alva Hodgson, who was to develop a regional reputation as a miller and millwright and whose surname graces the present structure,

Plate 12. Hodgson (Aid-Hodgson) Mill, Ozark County

purchased the original mill and operated it for thirteen years. In 1897 he rebuilt and expanded the original, replacing it with a three-story frame structure that remains today. Like most other water mills in the Ozarks, it was built using mortise-joint construction and contains much hand-hewn lumber, in this case mostly pine, which Hodgson reportedly cut and milled himself. Huge support beams were joined by tenons inserted into mortices and locked into place with a wooden pin or treenail. The next year Hodgson's brother George acquired part ownership in the enterprise and became its operator.

The Hodgson Mill was powered by two underwater turbines driven by a powerful and reliable head of water from the spring. Its millstones were reported to be imported French buhrstones from the Pyrenees. It was principally a custom mill that produced corn meal and flour primarily for local residents for a toll in kind. However, as one of several water mills in Ozark County (the Zanoni Mill, for example, is only five miles southwest on State Highway 181), the Hodgson Mill never monopolized milling in its area.

Alva Hodgson (who later built the Dawt Mill in 1899 and operated it until 1909) and his brother eventually disposed of their interests in the property. In 1934, Charles T. Aid acquired the mill. For that reason it is sometimes called the Aid-Hodgson Mill. For more than a century it produced meal and flour for the local residents of Ozark County. As late as 1965, for two cents per pound, one could have corn or wheat custom-ground into meal or flour.

Like so many of the old water mills in the Missouri Ozarks, the Hodgson Mill is located in an extraordinarily beautiful natural setting. So picturesque is the scene that its photograph has appeared on postcards and calendars and in publications such as *National Geographic*. In recent years the mill, easily seen and accessible from State Highway 181, has been converted into a tourist attraction. As can be seen in Wells's painting, an undershot waterwheel, which turns gently in response to the water from the spring, has been prominently added solely for cosmetic purposes. Rental campsites are also available near the millpond and the musical branch which results from the overflow.

It takes little imagination to envision how the mill with its extinct general store (built about the same time as the mill but not included in the painting) was an important economic and social center for nearby residents during the ownership of Holeman, Hodgson, and Aid.[15]

Jolly (Isbell) Mill

Built with slave labor in 1837 on the eastern fringe of Newton County by Thomas Isbell and his son John, the Jolly Mill (originally the Isbell Mill) was first a distillery and then a buhrstone gristmill. It had an undershot wheel powered by water diverted from fast-running Capps Creek. A substantial structure, the three-story mill rested on a foundation of hand cut limestone slabs—laid without mortar—taken from a quarry across the creek in Barry County. Its support beams were reported to be hand-hewn logs and its siding to be pit-sawed boards. These boards or planks were produced by two men using a long rip saw, one working above the log and the other standing in a pit below the log. No iron nails were used. The logs and boards were mortised and held with handmade wooden pegs or treenails. Its French buhrstones were shipped up the Mississippi River and transported across southern Missouri by wagon into Newton County.

Located near a much-traveled road running southwest out of Spring-

Plate 13. Jolly (Isbell) Mill, Newton County

field to Newtonia, then northwest to Baxter Springs, Kansas, and Indian Territory, the mill became the heart of the village of Jollification (see map, p. 41). Travelers en route west often stopped there to replenish their supplies from its general stores and to repair equipment at its blacksmith shop. John Isbell, the sole owner of the distillery and mill after 1852, prospered from this westbound flow of emigrants.

But with the coming of the Civil War, Isbell vanished from Jollification, leaving the mill idle and neglected. On two occasions in 1862, Union and Confederate forces clashed in the village without destroying the mill. Thereafter, marauders also spared it, although they burned all other buildings. The mill miraculously survived the war. Afterwards, George Isbell, a cousin of John, acquired the property for two hundred dollars at a sheriff's sale.

With the return of peace, Jollification slowly revived as settlers again moved west to Kansas and Indian Territory. Unfortunately for the town, the construction of the nearby Frisco railroad substantially diverted this renewed traffic and forced the village into decline. By the mid-1870s, Jollification was called simply Jolly.

At some point in the 1870s, George Isbell shut down his distillery and limited his business solely to milling meal and flour. According to local lore, he took this action because he did not want to pay a new federal excise tax levied on whiskey. He thereafter enlarged the millpond and strengthened the original milldam with additional sandstone blocks. In 1894, Isbell sold the works to George I. Brown, a nephew. Brown immediately sold the property to new owners, who lengthened the stone dam, further enlarged the millpond, replaced the undershot wheel with an underwater turbine, and installed rollers to supplement the millstones. The mill gradually became known as the Jolly Mill.

Until 1912, A. C. Lucas and Son operated the works as the Jolly Rolling Mills. When later acquired by W. F. Haskins, the Jolly Mill underwent substantial upgrading with the addition of a second turbine. After World War I, production diminished until it ended in 1973.

Listed on the National Register of Historic Places since 1983, the Jolly Mill is now being restored by the Jolly Mill Park Foundation. (Wells's painting of the mill was done before this restoration.) A new milldam has been constructed, the mill has been substantially rebuilt, and parking and picnicking facilities have been added. With these improvements the mill has taken on a new life as a tourist attraction. Although corn and wheat are no longer ground there, the mill is one of the few in Newton County that has weathered the storms of time.[16]

✻ McDowell Mill ✻

Water mills were scattered throughout the Missouri Ozarks wherever consistently reliable streams, creeks, or springs were found. The presence of such large numbers of mills in the late nineteenth and early twentieth centuries suggests their importance to the early inhabitants, who depended upon them for essential flour and meal. In the absence of the water mills, they were forced to rely on crude and primitive hand-mills or animal-powered mills for grinding corn and wheat into more edible products. Thus the building of a water mill such as the McDowell Mill was generally viewed as a godsend, by local residents who gladly used its services despite its distance and the great inconvenience of going to mill.

Flat Creek, which was fed by numerous springs, meanders through the heart of Barry County. More than a hundred years ago, a water mill was built on a branch of the creek about a mile down an unmarked road southeast of the present community of McDowell, the county seat for a

Plate 14. McDowell Mill, Barry County

while in the 1830s and 1840s. McDowell is located on County Road C approximately eight miles northeast of Purdy (see map, p. 41). Although the history of the original mill is lost, there is some indication that its location over Flat Creek subjected it to periodic flooding. Consequently, when Wilson, Issac, and John Hutchens acquired the property and rebuilt the mill, they switched the millsite to one that would be less prone to this hazard. They built the mill on a more elevated location some distance from the original branch site near Flat Creek. There it was somewhat protected, yet it was close enough to the rushing creek to draw upon its waterpower.

In either 1872 or 1886 (there is uncertainty concerning the date because sources differ), the Hutchens brothers constructed the present McDowell Mill. The three-story frame structure rested on a stone foundation. For the interior supporting beams the brothers chose oak, and for the exterior siding, pine. They continued to use the original stone dam, which was reinforced with concrete in 1915.

But because they selected a more protected location, the Hutchenses found it necessary to dig a millrace or canal and redirect water from Flat Creek into a forebay (the part of the race from which water was immediately released) that fed their underwater turbine. To transmit power from the turbine to their flour mill and sawmill located about two hundred feet away they installed a connecting steel shaft with a diameter of about three inches. By such means waterpower was used to turn the distant belt and pulley system driving the millstones. The newly built McDowell Mill was reported to be among the best in southwest Missouri, one that used modern rollers and added much to the local economy and the convenience of the McDowell community.

Like many of the old frame water mills, the McDowell Mill has succumbed to time. It enjoyed continuous operations for nearly seventy-five years before production ceased in the 1940s. As late as the 1950s, the structure was reported to be in good condition, with the site being frequently used for picnicking and fishing. Although in later years new owners made repairs that delayed the onslaught of time, the mill gradually deteriorated, as Wells's painting so graphically illustrates. Like the Dryknob Mill in Laclede County, it has collapsed into a sprawling mass of weathered timbers.[17]

✳ Noser Mill ✳

It would be difficult to find an old water mill that blends more harmoniously into its natural surroundings than the Noser Mill situated on the Bourbeuse River in central Franklin County. Located approximately fifteen miles to the southwest of Union, a stone's throw off State Highway 185 (see map, p. 41), the mill rises up from the river to merge beautifully with the foliage and landscape. Before recent modifications, this stone structure was a remarkable building that revealed the extraordinary skill and craftsmanship of its nineteenth-century builders. Even the modern changes do not hide the beauty of the clearly defined original structure, the back of which can be easily seen from a nearby one-way bridge over the Bourbeuse. Wells's watercolor captures the stately old mill as it was before the modifications.

Like so many of Missouri's early millwrights, the builders of the Noser Mill immigrated into the state from elsewhere. Dietrich Voss, a German immigrant, acquired land along the Bourbeuse River in 1849.

Plate 15. Noser Mill, Franklin County

Sometime during the 1850s Voss decided to exploit the power potential of the river by building a gristmill on his property. Employing local craftsmen, Voss erected a three-story stone structure that projected directly up from the water's edge on the south bank. He also built a stone and masonry dam which angled upstream away from the mill toward the north bank, thereby channeling the river toward the mill and its underwater turbines. The stonemasons who built this magnificent structure were reportedly paid fifty cents per day for their labor, while other workers were paid only twenty-five cents. Nonetheless, their work was superbly done, for after more than a century their craftsmanship, although recently altered, remains evident.

After the mill had been in operation for nearly two decades, John J. Noser and a partner purchased the property in 1871. Before the end of the decade, however, Noser bought out the interest of his partner and became the sole owner of the mill. There is little information concerning its operation during the Noser ownership; that is, little concerning production records, kinds of products (for example, meal, flour, mixed feeds), and so on. The size of the mill, however, suggests that it was fully automated and that its production of meal and flour was significant. It undoubtedly went beyond mere custom grinding for the local inhabitants and produced for more distant markets. Sometime after

Noser acquired full ownership of the mill, his son Edward A. Noser assumed responsibility for its operations. The son continued to manage the business and perhaps later owned it, until 1917. During the proprietorship of the Nosers, the post office at Luther was moved to the millsite, and John J. Noser became postmaster. In September 1901, the official name of the post office was changed to Noser Mill.

Sometime between 1917 and the late 1930s, the Noser Mill ceased producing flour and meal. However, during the 1940s the mill was used for generating electricity. The events of the last two decades have radically altered the appearance of the Noser Mill. In April, 1974, fire damaged the structure. Furthermore, in an effort to capitalize upon the tourist trade, its new owners have added a deck on the west side and added suites of rooms to the roof area that sharply changed the outline of the building. Nevertheless, the original structure is clearly discernible despite these modern innovations, and the Noser Mill remains an impressive structure. It and the home of the miller have been nominated for the National Register of Historic Places.[18]

✶ Old Appleton (McLain) Mill ✶

Settlers moving onto the eastern edge of the Ozark Plateau in the old Spanish Cape Girardeau District during the eighteenth and early nineteenth centuries were quick to see the waterpower potential in streams such as Apple Creek, which today separates Cape Girardeau and Perry counties. Like the Whitewater River and other streams in the vicinity, Apple Creek's volume was sufficiently constant to make it reliable and dependable for water-mill operations. It remained thus until the twentieth century, when men altered its watershed and diverted or reduced the flow of its natural tributaries. The volume of water in Apple Creek then became less predictable, a fact often reflected in the reduced operations of the Old Appleton Mill when the creek fell below the minimal level for uninterrupted milling. Before these disruptive changes in the watershed, however, snakelike Apple Creek offered a number of ideal millsites.

One of these sites was located about two miles slightly northwest and

Plate 16A. Old Appleton (McLain) Mill, Perry County

upstream from where the Old Appleton Mill was later built. Very little is known about the Kimmel Mill, which was constructed there in the 1740s. Like that of so many frontier settlements, the history of the mill and the original town of Apple Creek, where the mill was situated, is hidden in obscurity. The mill burned about 1790 and was not replaced.

About 1808 a new settlement began developing approximately two miles to the southeast on the creek's south bank in what is now Cape Girardeau County. Some of the settlers, however, elected to locate on the north bank in what is today Perry County. This general settlement was called Apple Creek until its southern part was incorporated as Appleton. The community was eventually called Old Appleton in order to differentiate it from Appleton City. Thus, when Old Appleton Mill was constructed by Alfred McLain in 1824, unlike many other water mills that gave birth to small communities, it became part of an existing settlement. Although controversy exists concerning whether McLain's two-story mill with its log dam was built on the south bank (Cape Girardeau County) or the north bank (Perry County), its location today is on the latter site. With its construction many residents moved nearer to the mill, an action that demonstrated its importance to the local inhabitants.

From 1824 until 1872, the Old Appleton Mill operated as a gristmill. It was driven by a waterwheel that turned French buhrstones to grind corn meal and a low-quality flour. Power from the wheel was also used to drive a sawmill that used a vertical or perpendicular blade in milling timber. During nearly half a century of operations, few improvements occurred to the mill. Under the ownership of James W. McLane, however, the mill underwent changes that incorporated new power technologies into its operations.

In 1872, shortly before disposing of the mill to Charles Hesse, McLane replaced the traditional waterwheel with underwater turbines that were more efficient and less subject to surface freezing. There were no additional changes in power generation until 1936, when Joseph Buchheit and Leo F. Unterreiner, co-owners from 1909 until 1949, replaced a turbine. Like many of the smaller water mills of the era, the Old Appleton Mill custom ground corn or wheat on demand, charging a toll which varied over the years. If the customer preferred, meal or flour in inventory was exchanged for corn or wheat, with an adjustment for the toll.

Many owners possessed the Old Appleton Mill during the course of its history. Although Charles Hesse added a third floor about 1891, twentieth-century owners were responsible for the principal alterations in the mill complex. When Theodore W. Meyer became owner in 1905, he replaced the log dam with one of concrete, only to have his work greatly damaged by floating trees. With the aid of Joseph Buchheit, a partner acquired in 1906, Meyer rebuilt the dam, which remains fixed in Apple Creek. (The top of the dam can be seen in Wells's painting of the mill from across the creek.) To increase the force of the water against the turbines and to add strength and stability, the dam was angled upstream toward the south bank of the creek.

In addition to changes in the milldam, alterations occurred in the mill itself. Either before or during the ownership of Theodore Meyer, rollers were added to improve the quality of the flour. During the ownership of Buchheit and Unterreiner, the sawmill was detached from the west side of the mill and moved to a separate site about fifty yards north. A hammermill was added and enclosed where the sawmill had been located. In addition, a cupola was placed on the roof, a new millrace was constructed, and a new stone and concrete wall was built to enclose and protect the turbines. Furthermore, when Arthur Schulze acquired the mill in 1949, he attempted to elevate the first floor against floods, a move which forced changes in the frontal loading dock. During the twentieth century, therefore, the owners of the Old Appleton Mill altered its outward structure from that seen in Wells's two paintings.

Like many of its counterparts throughout Missouri, the Old Appleton Mill fell victim to forces beyond its control. Production declined steadily from 1936 until it ceased in 1948. Competition from larger and more

Plate 16B. Old Appleton (McLain) Mill, Perry County

efficient mills, alterations in the natural watershed which created water shortages, and the greater mobility of the population forced an end to milling. Although ownership of the mill has changed several times since 1948, not much was done to preserve the property until it was acquired by Art and Rene Dellamano in 1977.

Unfortunately, the Dellamanos' efforts fell victim to that old nemesis of water mills—the flood. In May, 1986, extremely high water in Apple Creek swept the mill off its foundation and pounded it into pieces. This natural disaster was a dramatic ending to a cultural landmark that linked a declining community to its past. When crossing the bridge over Apple Creek on U.S. Highway 61 at Old Appleton (see map, p. 41), travelers now see only the foundation and dam of the Old Appleton Mill.[19]

* Ritchey Mill *

Many sites for water mills were located on the banks of the swift flowing streams of Newton County. During the course of the nineteenth century, many small entrepreneurs recognized the tremendous power potential in such creeks as Shoal and Capps. Consequently, this Ozark county came to have its share of the state's early water mills, several of which have survived long after their heyday. One of these holdouts against time is the Matthew Ritchey Mill.

   Located about five miles to the northeast of Granby off U.S. Highway 60, the Ritchey Mill sits on the south bank of beautiful Shoal Creek less than a mile south from the decaying town of Ritchey (see map, p. 41). Fed by an estimated twelve hundred springs upstream between Ritchey and Pioneer, Shoal Creek was generally characterized by great volume and rapid flow except in times of drought, as in the 1930s. An unpainted two-story frame structure, the Ritchey Mill occupies the same site as the original that was built in 1841.

Colonel Matthew Ritchey owned the first mill, using slave labor to construct it and the dam across the creek. About the same period, Ritchey also used slaves to build two mansions, one in Newtonia and the other in Ritchey. The original mill, a gristmill, was powered by a vertical, twelve-foot undershot waterwheel because of the location and the volume of water flowing in Shoal Creek. It used millstones for grinding meal. These stones can still be seen today at the mill site. Building Ritchey's milldam in the 1840s was a major accomplishment. It consisted of logs, hand-hewn timbers, and stones. Like so many of the early milldams, boxed cribbing was skillfully used: hand-hewn square timbers were placed across the stream flow, round logs were fixed parallel to the flow, and the cribbed boxes were filled with stones. With occasional repairs, this original dam remained until it was redone after World War II.

Although the Ritchey Mill was in the vicinity of Civil War battles and skirmishes, fortunately it was not destroyed, as were so many other mills, by contending Union and Confederate military forces. From its building until 1867, Ritchey retained exclusive control over the mill. But in 1867, he and seven others, who had invested $1,500, incorporated themselves into the first Ritchey Milling Company. This company lasted

Plate 17. Ritchey Mill, Newton County

until the turn of the century, when the original mill was replaced and the company was reorganized.

Sometime between 1910 and 1912, underwater turbines manufactured in New York were installed to replace the waterwheel. Nevertheless the mill remained a gristmill that produced only corn meal. Under the new company, an icehouse was added to the rear. During the winter eight- and ten-inch blocks of ice were cut from the pure waters of Shoal Creek and stored in the icehouse for sale during the summer. Manford Troxel, an old-timer of Ritchey, recalled that the mill did a thriving ice business throughout the surrounding area.

After fire destroyed the second mill in 1931, the present mill was built the following year. As Wells's painting from across Shoal Creek beautifully illustrates, it is an impressive structure. Rollers were installed that made possible the production of a high-quality flour. Because the drought of the thirties made Shoal Creek unreliable as a source of power, a powerful diesel engine was brought from Springfield, Missouri, and installed. Like its predecessors, the new mill continued to serve the community of Ritchey and the surrounding area, providing meal, flour, feeds, ice, and, for a while, even electricity. Old-timers remember stories of fifteen to twenty teams and wagons loaded with white corn waiting in line to be loaded or unloaded at the mill. They also fondly recall the high-quality meal and flour bearing the Shoal Creek Mill brand name that was shipped throughout southwestern Missouri.

The town of Ritchey did not result from the mill. It originated during the construction of the Atlantic and Pacific Railroad, which became a part of the Frisco line. However, the nearby presence of the mill clearly contributed to the economic well-being of the community, which thrived during the first two decades of this century. The waters behind the milldam were used for fishing, swimming, skating, baptizing, and picnicking. Regular production in the mill ceased in 1958, and it was totally shut down two years later.

Like that of most surviving mills, the ownership of the Ritchey Mill has changed many times. In 1959 the property was acquired by G. E. Thomas, who intended to renew regular production of meal and flour. Thomas canceled his plans, however, because of the expense of bringing the works into compliance with federal regulations. Now owned by Gorton Thomas, the mill with all its machinery has remained silent for more than twenty-five years, apparently waiting for someone to restore it to its glory days.[20]

❋ *Schlicht Mill* ❋

Like so many of Missouri's early water mills, the Schlicht Mill was built by a migrating Southerner. Joseph Strain, a Tennessean who moved west into Missouri along the Kickapoo Indian Trail, found a substantial spring in an isolated hollow near present-day Crocker and Swedeborg (see map, p. 41). A craftsman who owned his own millwright tools, Strain saw the potential of the site for a gristmill. Not only was there a sufficient quantity of water to drive the mill, but there was also a bountiful supply of oaks, sycamores, and limestone for its construction.

Beginning in 1840, Strain began the tough work of cutting limestone blocks from nearby bluffs for the foundation and dam and hewing the huge oak support beams needed to undergird the mill. Four years later the mill and dam were completed. Overflow from the pond ran into the nearby Gasconade River, and consequently, although owned by Strain the mill was locally known as the Gasconade Mill. It is reported that in 1862 Strain sold his mill properties to John Hensley in order to return to

Plate 18. Schlicht Mill, Pulaski County

Tennessee and participate in the Civil War. Fourteen years later (1876), Hensley sold out to John Schlicht, who had immigrated to the United States from Frankfurt, Germany, in 1866.

Schlicht, whose parents and brother joined him in 1870, was a miller who worked for a while in New England and New York before moving west to Lebanon, Missouri. There he managed the J. F. Smith Mills for six years before acquiring the Gasconade Mill. Like many others who gained ownership of older mills late in the nineteenth century, Schlicht set about making major improvements in order to serve potential customers who lived within a radius of fifty miles of the mill. Among the changes was a second millpond to increase the head of water for two additional waterwheels. To supplement the power of the mill, Schlicht installed a steam engine and thus became one of the earliest Missouri millers to combine water and steam power successfully in operating a mill. Furthermore, he installed rollers so that he could produce flour as well as meal, and he reportedly developed a new process for bleaching flour. To protect his mill from encroachments, he purchased additional land, bringing his total holdings to approximately two hundred acres.

Like many other mills of the era, the Schlicht Mill became a kind of trading and social center for the farmers who brought their grain by mule and wagon from distances of forty to fifty miles. Upon their arrival, provisions were made for feeding the animals, and, if an overnight stay was necessary, the farmers usually slept in the mill with their grain. Near the mill Schlicht set up a country store, a barber shop, and a tavern. The whiskey sold in the tavern was produced in the Schlicht home.

Before the end of the century, the Schlicht Mill also provided postal services after the Frisco Railroad established a passenger stop approximately a mile from the mill. There is some indication that with improved transportation and the building of a clubhouse (available for social events) on the pond nearest the mill, the number of people visiting the millsite greatly increased. The attractiveness of the surroundings is evident in Wells's across-the-pond painting of the mill.

The Schlicht mill was operational for nearly a century. When Charles Schlicht, who had inherited the property from his father, John, died in 1945, the property was tranferred to Dr. William Schlicht, a nephew of Charles, who set about restoring and preserving the mill properties. Sherman Schlicht later acquired ownership and retained possession until 1976, when he sold the mill to a Saint Louis investor. Unfortunately, the mill was allowed to deteriorate and has now collapsed. Its equipment has been removed and used to restore the Bollinger, Dillard, and Montauk mills operated by the state.[21]

Topaz Mill

In its heyday between 1895 and the late 1930s, the Topaz Mill was the center of a thriving community in Douglas County. In addition to the mill, there was a general store, a blacksmith shop, a cannery, a barbershop, and a post office. And before the founder and owner of all these enterprises died in 1919, there were even plans to establish a bank. Around the mill, therefore, was the embryo of the small town of Topaz, which, unfortunately, died aborning. Like so many small villages that sprang up in isolated places because of water mills, Topaz fell victim to external developments.

Located about fourteen miles southwest of Cabool near the North Fork of the White River, the Topaz Mill was situated near a spring that continues to produce approximately eight to ten million gallons of water each day (see map, p. 41). It was not, however, the first mill to be fed by this reliable source of power. Sometime during the late 1830s a Choctaw Indian woman named Alabath Freeman obtained ownership of the land

Plate 19. Topaz Mill, Douglas County

on which the spring was located. About 1840 she and her husband Aaron moved with his children and their families from Carter County and settled near the spring. There they reportedly built the first small water mill, a gristmill, which was powered by the rapidly flowing water. Unlike the present Topaz Mill, the Freeman mill was located adjacent to the spring. Between the ownership of the Freemans, who died in 1861 and 1862, and Robert Samuel Hutcheson, who acquired the property in 1890, the property changed hands numerous times. The spring, however, continued to power a small gristmill.

Robert S. Hutcheson and his brother moved into Douglas County in 1890 and bought forty acres of land on which the spring was located. Three years later the brother sold out his interest to Robert, who gradually accumulated about eighteen hundred acres in the vicinity. Hutcheson saw the great potential of the spring and in 1895 employed a local carpenter named Smith to construct the three-story Topaz Mill. Because a fully automated mill was planned—one to produce flour as well as meal—it is likely that the manufacturer of the milling machinery sent his own millwright to work with the carpenter in integrating structure and machines.

For such a mill it became necessary to increase the power from the spring. Consequently, a dam was built, and the water from the spring-fed millpond was redirected to the works by a wooden flume to drive the turbines. The mill had a top capacity of about forty barrels of flour per day, each barrel weighing ninety-six pounds. It was, therefore, a "forty barrel mill," a combination "custom" and "merchant" mill that ably served the local population. Wells's painting is a faithful and appealing rendition of the magnificent old structure.

Most of the mill's auxilliary developments occurred during the ownership of Hutcheson. Although a John Talley had earlier operated a log cabin store, it was Hutcheson who built a substantial general store, which still stands, in 1913. This structure matched the mill in beauty, having a balcony; showcases and shelves containing ladies' and men's wear, dry goods, and hardware; and storage for produce in its cool basement. A cannery was added to preserve tomatoes and other vegetables. A blacksmith shop took care of local horseshoeing and iron needs. In a sideroom off the mill, the miller operated a barber shop. And inside the general store, Hutcheson's wife Mary operated a post office. Before his death in 1919, he had planned to establish a bank near the mill.

After Hutcheson's death, his wife and children operated the businesses until 1945. However, milling operations gradually subsided and eventually ceased. Flour production ended in the early thirties; meal production stopped in the late thirties. The general store remained open until 1945, when the wife retired as postmistress and sold the mill properties at public auction to Bill Southerland, a son-in-law. Two years later Southerland sold out to Noble Barker, who did not operate any of the businesses associated with the mill. Thus, all operations ceased. Joe W. O'Neal, the present owner, who purchased the property from Barker, now operates a handle factory in the old general store while farming and restoring the mill.[22]

✳ Zanoni Mill ✳

Situated off State Highway 181 approximately five miles to the southwest of the Hodgson Mill near Sycamore in Ozark County, the Zanoni Mill remains intact with its overshot waterwheel in place (see map, p. 41). Although the area surrounding the millsite has been substantially altered to accommodate a large modern dwelling, the basic structures of the mill and a nearby general store have been remarkably preserved and can be easily seen from Highway 181. Despite the alterations in the physical environment, which have made the mill and store mere showpieces of the dwelling, they continue to retain the dignity of olden days, when they were true centerpieces of a functioning community.

Like its companion Hodgson Mill to the northeast, the Zanoni Mill derived its power from one of the many springs found in the Ozarks. However, the daily volume of the Zanoni Spring was considerably less than that of the Hodgson and other springs, the former being only 200,000 gallons and the latter about 24,000,000. Furthermore, the locations of the springs were different. The Zanoni Spring overflowed from nearby bluffs, and the Hodgson spring emerged at the base of a bluff. The small volume and the elevated location of the Zanoni Spring determined, as in the case of the Falling Spring Mill, that any mill built at the site would necessarily be small. Thus, unlike many large and fully automated water mills built on large creeks or rivers (for example, the Bollinger, the Noser, the Dawt, and the Old Appleton mills), the Zanoni Mill was one of limited size and capacity.

There have been three water mills located on the Zanoni millsite. The first was constructed of logs. It was known to exist during the Civil War, which caused the destruction or closing of many Missouri mills. Little is known, however, about the builder and owner of this original gristmill. The primitive mill undoubtedly served the needs of local residents, who, because of the difficulty of moving about in the Ozarks, found it more convenient for their milling than the nearby Hodgson mill (which may have been shut down during the war). It appears nevertheless that a mill continued to operate at the Zanoni Spring site until 1900, when John Cody and George Shoemaker built a new gristmill there. Unfortunately, the life of their mill was short. Within five years it had burned, making possible the construction of a third mill.

Shortly after the burning of the second mill, A. P. Morrison acquired

Plate 20A. Zanoni Mill, Ozark County

the millsite and built the present Zanoni Mill, a small, odd-shaped, two-story frame building that he positioned close to the bluffs in order to use the spring. A flume was constructed to redirect the springwater over and onto an external waterwheel. Thus the new mill was powered by an overshot wheel turned by a spring of small but reliable volume. These features are portrayed beautifully in Wells's two paintings. Sometime during the 1940s, it became necessary to replace the wheel with a new one.

Morrison was an entrepreneur who was not satisfied to operate a small gristmill. Records indicate that during his ownership several small businesses developed around the mill, forming the tiny village of Zanoni. Shortly after rebuilding the mill, Morrison established a cotton gin that operated with power from the waterwheel. He also built a general store, a small building that remains near the mill and is seen in one of Wells's paintings, and a blacksmith shop to service the teams that hauled grain for grinding. In 1920 he created a small overalls factory. Much of the business life of Zanoni resulted from Morrison's efforts. In addition, he also served for years as postmaster.

Following World War II, the nationwide developments in milling and transportation and the greater mobility of the local population under-

Plate 20B. Zanoni Mill, Ozark County

mined the need for small gristmills like the Zanoni. Efforts were made to find new uses for them. Renovations were made on the Zanoni Mill in the forties so that it generated enough electricity to supply two dwellings and a country store. At the same time, it continued to function as a gristmill, grinding out small quantities of meal for local farmers. These efforts failed to save the mill, and it closed in 1951.[23]

* *Epilogue: Going to Mill* *

The water mills described and portrayed in this volume represent only a tiny fragment of the scores that flourished on or near Missouri's rivers, creeks, and springs at the turn of the century. But even then the number of mills had been dramatically reduced by the conversion of many of them from water to more efficient forms of power and by the development of more sophisticated methods of milling. Just as the waterwheels, those primitive symbols of a developing industrialism, were replaced by underwater turbines at many mills, so in time the turbines were replaced by steam engines which in turn were replaced by electricity and diesel engines or a combination of both. These increasing switchovers to newer power technologies enormously increased the efficiency and productivity of the mills in which they occurred. At the same time, they steadily reduced the need for the marginal water mills that used one obsolete form or another of waterwheels to drive their millstones or rollers. In much the same fashion, by the turn of the century improved transportation and the growth of modern milling, which tended to concentrate in the state's urban areas, such as Cape Girardeau, steadily reduced the need for the water mills that had flourished earlier in isolated areas of the Ozarks.

This revolution in power and milling systematically drove the primitive water mill, with its slowly turning wheel, into extinction, unless, of course, it happened to be situated somewhere in an isolated hollow that offered temporary haven against ever advancing technological development. Infrequently in such places a water mill can yet be found that partially fulfills the lingering, nostalgic image of these dead workhorses of the past. The Zanoni is such a mill. When viewing it, one can only speculate about the vital role that the old mill and its hundreds of counterparts throughout the state played in the social and economic life of the communities they helped to create and sustain.

Far more likely than not, however, the complete history of an existing mill has been either lost or so entangled with local lore and fiction that it is unrecoverable. In their heyday, primitive water mills were so commonplace throughout the state that no one thought it important to preserve their individual histories. Very few contemporaries anticipated the near total demise of the mills. If they had done so, perhaps more complete records would have been preserved.

The men who built, owned, and operated the early water mills in Missouri—even before it became a state—were a special breed whom today we call entrepreneurs. They were tough, enterprising men who courageously entered the trans-Mississippi wilderness with a willingness to risk their lives and fortunes on speculative ventures. Of the water mills included in this volume, many were built and owned by migrants from the Southern states and elsewhere. As the historical vignettes clearly reveal, builders like George Frederick Bollinger (Bollinger Mill), Mathias Bollinger (Dolle Mill), Samuel Greer (Greer Mill), Joseph Strain (Schlicht Mill), and Alva Hodgson (Dawt and Hodgson mills) were obviously a "cut above the average" in intelligence, physical stamina, and resourcefulness. They were men of vision who saw the need for more sophisticated milling to save the state's pioneer and rural population from relentless drudgery. They used their talents and resources to meet that need.

The Bollingers, the Greers, and the Hodgsons were always welcomed into a frontier region. As millwrights, carpenters, and farmers, they aided tremendously in subduing the wilderness environment. They nearly always engaged in other occupations as well as milling, frequently combining it with farming or some other trade, because the mills they established generally provided them with only a partial living. Nevertheless, because of their talent for leadership and their significant role in economic development, they frequently emerged as prominent persons in the communities they helped to establish. In many cases their families came to dominate the economic life of the area, extending the initial influence of the mill founder.

Typical of the men who helped to establish Missouri's milling industry in the early days was M. F. Byrne, an early owner of the Brynesville Mill in Jefferson County. A contemporary said of him: "There was that about him, . . . which always wins success. He was courageous, and daring, quick to enter on any enterprise and with perseverance enough to see it through in the face of all obstacles. He was industrious and energetic. He possessed a certain sharpness and shrewdness, but was strictly honest, and was known as a business man of straight-forward and honorable dealing."[1] Although Byrne did not establish the Byrnesville Mill on the Big River, the qualities attributed to him were undoubtedly similar to those possessed by scores of men who penetrated Missouri's backcountry, discovering and harnessing the potential power in its numerous rivers, creeks, and springs.

Later owners of the mills were equally ingenious as entrepreneurs. Lester Klemme, who acquired ownership of the Dillard Mill in the 1930s during the Great Depression, continued to operate it as a custom mill, that is, grinding wheat and corn on demand for his customers. Perhaps responding to the hard times, he sought to use the mill in innovative ways to increase his income. In addition to producing flour and corn meal, with a homemade mixer he began to grind and mix feeds for the livestock of his patrons. Aware of the natural beauty of the millsite, he also built several small cabins and a large farmhouse which he operated as the Old Mill Lodge. Connecting a generator to the turbines of the mill, Klemme lighted his lodge with electricity. His out-of-town guests enjoyed the scenery, swam in his millpond, read late into the night, and ate homecooked meals prepared by his wife—all for a fee, of course.[2]

The men who built the water mills included in this volume were a remarkable group, but, as noted earlier, the mills of Missouri did not originate in the inventive genius of these entrepreneurs. By the time they arrived, the water mill had already proved itself as an advance agent of civilization in both the British North American colonies and later the early American republic. The presence of a water mill and millpond in an area was a striking harbinger of the end of the frontier, a mark of stability and permanence. Through the Bollingers, the Greers, the Strains, and others, early Missourians became the beneficiaries of nearly two hundred years of American experience in building and operating water mills.

Opening new lands in Missouri was a back-breaking task and, consequently, labor saving devices of any kind were enthusiastically welcomed. The building of a water mill, either a sawmill or a gristmill, nearly always soon followed the penetration of a region like the Missouri Ozarks that had abundant water and timber.

Providing food and shelter, of course, was a necessity that consumed much time and hard labor. The preparation of food, particularly bread, which was an essential of frontier diet, was an especially arduous task for women. Using primitive hand mills like the mortar and pestle or the quern, a frontier woman devoted much of her time in dull, monotonous labor in order to prepare the corn meal for her family's daily bread. And if her home was constructed of anything other than round or flat logs hewn and notched with an ax, it was because her men torturously sawed boards with a two-man pit saw. For these reasons, the early settlers of Missouri encouraged the construction of water mills that helped to reduce the drudgery of providing food and shelter. The water mills portrayed in this volume frequently powered both the sawmill and gristmill in combination.

Although establishing a water mill was often the start of a community (for example, the villages of Topaz and Drynob), excellent millsites were often situated in such isolated places that access was extremely difficult. To illustrate, the original Greer Mill was located on a nearly inaccessible site close to an extraordinary spring which provided its power. Consequently, no village or town developed around it. Furthermore, in the early days potential users of a water mill often lived great distances away. Regardless of the difficulty of access, the need for the services of the mills was so great that "going to mill" became a regular practice for patrons who might live many miles away.

In doing so, early Missourians duplicated the experiences of eastern settlers, who willingly traveled great distances in order to use a gristmill. Contemporary accounts indicate how important such mills were to early Americans living east of the Mississippi River. An unidentified Pennsylvanian wrote about his trip to mill around 1810: "I had fourteen miles to go in winter to mill with an ox-team. The weather was cold and the snow deep; no roads were broken, and no bridges built across the streams. I had to wade the streams and carry the bags on my back. The ice was frozen to my coat as heavy as a bushel of corn. I worked hard all day and got only seven miles the first night, when I chained my team to a tree, and walked three miles to house myself. At the second night I reached the mill." And later, Horace M. Kephart wrote the following about a mountaineer neighbor in Appalachia: "Before there was a tub-mill [a primitive form of water mill (see Fig. 9)] in our settlement one of my neighbors used to go, every other week, thirteen miles to mill, carrying a two-bushel sack of corn (112 pounds) and returning with his meal on the following day. This was done without any pack-strap but simply shifting the load from one shoulder to the other, betimes."[3]

These experiences were not too unlike that of A. O. Weaver, a Missourian, who used the Cedar Grove Mill in the early part of this century. Weaver lived some distance from the gristmill that he felt compelled to visit on a regular basis. In describing his going to mill, he stated: "We've got to have corn meal at our house or we can't live. I've got a big family an' it takes lots ov bread, ... This ol' Cedar Grove Mill is a real ol' timer an' has been grindin' out corn meal ever since before the Civil War. It has purtnye raised my family 'cause there is where I've allers took my corn to have it made into meal, an' it's the kind we all like, 'cause it's ground on the ol' French burrs."[4] Placing two sacks of corn across the rump of his white mule, Weaver took his gun, dogs, and fishing equipment, straddled the animal, and went off to mill (Fig. 11). While waiting for his corn to be ground, he used the time to hunt squirrels and fish in order to supplement his family's food supply. His periodic treks to the water-powered Cedar Grove gristmill were a common experience for many rural Missourians in the nineteenth century—and even into the early twentieth century. Before the proliferation of such mills, journeys

Fig. 11. Going to Mill

of ten to fifteen miles to mill on foot, mule, or wagon were not uncommon for country folk in the Ozarks and elsewhere throughout the state.

With the proliferation of water mills in Missouri during the mid-nineteenth century, however, access to them became less of a problem except in remote areas. They tended to be closer together, thereby reducing the distance a farmer had to travel with his corn or wheat and providing him, in some cases, with a choice. Some counties often had a number of mills in relative proximity to each other. Ozark County in south central Missouri was such a county. For example, its Hodgson and Zanoni mills, included in this volume, were located about five miles apart.

Indications are that approximately 700 mills were in Missouri in 1840, nearly 400 in 1860, about 850 in 1870, and about 900 in 1880.[5] In the period after the Civil War, of course, many of these were not water powered mills. Nevertheless, the large numbers of mills reflect their extreme importance to Missouri's growing rural population. Until a network of roads opened up the backcountry to the flour and meal products of the large urban milling companies, rural Missourians continued to rely on the small "custom" mills for their grinding needs.

Although frequently built in isolated locations, many of the early water mills contributed to the creation and growth of small communities.

Going to mill was important in other ways than just to reduce the drudgery of the preparation of flour and corn meal. Milling days brought together in one spot widely scattered neighbors who were starved for companionship and news. Congregating around the mill while waiting their turn, they enjoyed each other's company and exchanged bits and pieces of information about deaths, marriages, land sales, newcomers, politics, religion, scandals, and so on. Some, like A. O. Weaver, mentioned above, used the waiting time to supplement the family's food supply by hunting and fishing (Fig. 12). Others who had come great distances often found so many of their neighbors ahead of them that they had to wait overnight to mill their grain. It was not uncommon for the mill owner, as at the Schlicht Mill, to provide his customers campsites and feed for their mules, horses, or oxen.

Enterprising millers and others saw the financial potential inherent in the needs of the people in these encampments. Subsidiary businesses sprang up to supply those needs. Lumber was nearly always in demand, a demand which could be met with a sawmill powered by belts and pulleys connected to the principal shaft of the gristmill. A number of the water mills in this study (for example, the Old Appleton Mill and the Drynob Mill) combined gristmilling and sawmilling. Furthermore, in nearly every case a blacksmith shop was established both to maintain

the mill's machinery and to provide shoeing services for the draft animals of local famers. Manford Troxel, an old-timer who lived all his life in Ritchey, recalled that often the waiting line of mules and wagons loaded with grain for the Ritchey Mill was a half-mile long or longer, backed up from the mill across the bridge over Shoal Creek. To prevent a farmer from losing his turn, the local blacksmith moved busily along the line, replacing horseshoes where needed. Troxel observed that at a time when the primary source of mobile power was the draft animal, access to a smithy's services when going to mill was an enormous convenience to farmers.[6]

But it was the establishment of the general store or stores near the mill that marked the settlement's development into town or village. Supplying dry goods, hardware, and groceries, the stores made going to mill even more attractive to farm families who lived widely separated from each other. With the addition of mercantile establishments and residential dwellings to the cluster surrounding the mill, "going to mill" now meant "going to town." The Bollinger, Byrnesville, Cedar Hill, Dawt, Drynob, Hammond, Jolly, Noser, Old Appleton, Ritchey, Topaz, and Zanoni mills all had at least one or more general stores. For the times, some of them were housed in substantial buildings. Even today, for example, the interior of the store adjacent to the Topaz Mill reveals evidence of a

Fig. 12. Fishing While Waiting Turn

once thriving business. Its railed balcony, counters, shelving, and showcases—all still intact—suggested that the owner had run a prosperous enterprise.

More often than not, the general store was also the location of the post office, the presence of which was another indication of the changed status of the developing community around the mill (Fig. 13). With the addition of churches and professional persons—perhaps a doctor, a lawyer, or a banker—some communities showed every indication of developing into permanent towns. Hammond was such a community. Some of the towns (for example, Cedar Hill) grew to the point that their mills, although important to their economies, were not essential to their continued existence. In contrast, when developments forced the closing of the Byrnesville, Topaz, Zanoni, Drynob, and Hammond mills, the small villages that they had helped to spawn began to wither and die. But for a while, they had flourished.

Drynob, on the Osage Fork east of Lebanon, was typical of such towns. At one time it thrived. With the Drynob Mill its early focus of community life, the village prospered as a small established community providing the surrounding area with the services it needed. At times, with its yard filled with teams and wagons loaded with wheat and corn to be ground into flour and meal, the mill ran day and night to

Fig. 13. Receiving Mail at the Mill

meet the demand. Its sawmill supplied lumber to the newcomers that moved into the vicinity. For a long time a succession of blacksmiths operated two shops there. First one and then several general stores provided bolt cloth, overalls, pants, and other dry goods, together with hardware and grocery products needed by local farmers. A post office was early established and remained open continuously for eighty-one years, finally closing in 1957. Dr. Tom Casey moved into the village and practiced medicine there for many years. Congregations of Moravians and Free Will Baptists established themselves at Drynob. And the hall over one of the general stores served for a while as a meeting place for a chapter of the Odd Fellows lodge, the local grade school, and the Free Will Baptists. The mill was truly an agent of community. But as milling declined, so did Drynob.[7] In October, 1986, the Drynob Mill, which had long ago ceased operations, collapsed. Its fall was a symbolic end to a town whose early residents basked in its promise.

Just as the early water mills promoted the origin and growth of small villages, they also promoted recreational and social activities (Fig. 14). Most of the millponds were great for fishing. Gorton Thomas, the present owner of the Ritchey Mill on Shoal Creek, remembers stories from old-timers about the excellent fishing behind the dam near the mill. Catfish especially flourished on the refuse from the mill, reducing the supply of the more desirable fish like trout. Nevertheless, fish of all kind abounded and attracted fishermen from near and far.[8]

Good fishing and the beauty of the millsite attracted visitors from such places as Monett. Writing in 1952 about historic spots in Barry County, Nellie Alice Mills described her early excursions to the Ritchey Mill. She wrote:

> Ritchey was of special interest to fishermen from Monett, who could go down there on the train. The mill by Shoal Creek is about a quarter of a mile from the railway station. Many a picnic party from Monett has spent a pleasant day there by the mill. The trains ran just right to suit us. We could go down in the morning and return in the afternoon. There was a grove fringing the stream below the mill where there was plenty of space to spread a meal in the shade, while across the wagon bridge, a short walk took us to a wide open space where we could build a campfire to cook our fish. Under the wagon bridge the creek was just deep enough to make a fine bathing pond.[9]

In an era when recreational options were few, the millsite was clearly a favorite place, especially on Sunday afternoons, for the young people who knew its charm. In addition, Shoal Creek below the mill was the customary place for local churches to baptize their converts. It contin-

ued to be so until the early 1980s. According to Thomas, most of the elderly living in Ritchey were baptized in the creek below the mill.[10]

Baptizings, swimming, fishing, boating, picnicking, camping—all took place on Shoal Creek near the Ritchey Mill. The same was true, however, for most of the old water mills included in this volume. As life around them evolved, it became increasingly social, changing the place and the significance of the mills in the communities they served. Although they were not established for the purpose, they nevertheless greatly facilited social intercourse in rural Missouri. In doing so, they greatly enriched agrarian life, helping to make living in isolated areas more enjoyable.

In examining the fate of the water mills included in this volume, one can glimpse the destiny of many other old mills found in the Missouri Ozarks during the late nineteenth and early twentieth centuries. The Noser and the Byrnesville mills were substantial three-story structures with stone foundations and brick walls. In an attempt to commercialize the natural beauty of its setting, the present owner of the Noser Mill has renovated the building into apartments for tourists. Although the architectural changes are not appealing, they have extended the life of the old mill by allowing its continued use. Major changes have also been imposed on the Byrnesville Mill to convert it into a private residence.

Fig. 14. Playing Horseshoes at the Mill

The alterations thus far have been tastefully done, with much of the mill decor being retained in the interior. It promises to be a beautiful residence upon completion. The Zanoni Mill, with its general store, has also been refurbished and made into an exterior showpiece of a large modern residence.

Fate has been kind to the Alley Spring, Bollinger, Dillard, and Jolly mills. In each case, full restoration is in progress or has been completed. Alley Spring Mill has been restored and is operated by the National Park Service. Both the Dillard and Bollinger mills have been converted into state historic sites, with on-site administrators. Under the Missouri Department of Natural Resources, Division of Parks, Recreation, and Historic Preservation, the Dillard Mill has been fully restored and the Bollinger Mill restoration is nearly complete. Both are excellent examples of the sophisticated processes built into the water mills of yesterday. With the strong support of the Jolly Mill Park Foundation, the Jolly Mill has been completely renovated, and the improvements promise a bright future for the mill. These restorations provide numerous visitors an opportunity to see an important economic and social institution of the past. They also extend the life of the mills.

Four of the mills—Cedar Hill, Dawt, Dolle, and Hodgson—remain operational to some extent. Of the four, Cedar Hill Mill is the greatest producer. Although it developed a substantial ice-making sideline years ago, it now concentrates upon the production of mixed feeds. The Dolle Mill does only limited custom grinding of feeds. Although the Dawt and Hodgson mills operate, their production is principally sample products for visitors and tourists. Yet the continued operation of the mills assures their preservation.

Of all the hazards that threatened the mills, fire and flood were among the worst. However, neglect was a close runner-up. None of the mills included in this study was destroyed by fire. The Old Appleton Mill, which was in the process of being restored by private owners, was washed away in a great flood that destroyed everything but its foundation. It is no more. After years of neglect, the Drynob, Schlicht, and McDowell mills fell victims to the elements. After fighting desperately to stand for years, they simply collapsed when rotting support beams gave way.

The fate of the other mills—Falling Spring, Greer, Hammond, Ritchey, and Topaz—hangs in the balance, depending largely on the decisions of private owners. The proprietors of the Topaz and Ritchey mills have exercised a careful watchcare over their old structures, which are in excellent shape with a full complement of machinery. Under the present ownership, they are likely to last indefinitely. And the owner of Greer

Mill is vigilant about protecting his property from vandalism. However, Greer Mill, like the Hammond and the Falling Spring mills, shows evidence of neglect and the perils of weathering. The roof of Hammond Mill has fallen victim to the wind, and the flume of Falling Spring Mill has collapsed. It appears inevitable that, in time, these mills will suffer the fate of the Drynob.

    When viewing these old structures, one can easily understand why poets and artists continue to find them fascinating as subjects for their work. Hammond Mill stands defiant, daring the elements to finish it off. The Greer, Ritchey, and Dolle mills cry out for rediscovery. The Topaz, Falling Spring, and Zanoni mills exude a smug serenity, happy with their lot in old age. The Drynob, McDowell, Schlicht, and Old Appleton mills plead for remembrance of their glory days. The Bollinger, Dillard, Alley Spring, and Jolly mills throb with the joy of new beginnings. The Hodgson and Dawt mills struggle to retain their old images, while the Noser and the Brynesville mills wrestle with new identities. The Cedar Hill Mill grinds on, seemingly unaffected by the passage of time. All of them speak of a past when their role was different; all of them plead for sensitive souls to hear their message. Jake Wells, as his watercolors indicate, represents the artists and poets who have listened.

✱ Notes ✱

Introduction: The Legacy

1. O. B. Bunce, "Scenes on the Brandywine," in *Picturesque America; Or, The Land We Live In*, ed. William Cullen Bryant, Vol. I, pp. 220, 222.
2. Matt. 18:6.
3. Bunce, "Scenes on the Brandywine," p. 222.
4. For the ancient developments that climaxed in the revolving quern, the slave- and cattle-powered mills, see Richard Bennett and John Elton, *History of Corn Milling*, Vol. I, *Handmills, Slave and Cattle Mills*, *passim*; Vol. II, *Watermills and Windmills*, pp. 6–7, 9. See also Abbott Payson Usher, *A History of Mechanical Inventions*, rev. ed., p. 161.
5. Usher, *Mechanical Inventions*, pp. 167–68; Bennett and Elton, *Corn Milling*, II, 9–10.
6. Usher, *Mechanical Inventions*, p. 180; In his *Stronger Than a Hundred Men: A History of the Vertical Water Wheel*, p. 7, Terry S. Reynolds wrote: "For most of the era of water power, the typical horizontal watermill was capable of generating little more power than a donkey or horse, and often not that much. It could be used only for a single task (milling grain), and it was wasteful of water.

The horizontal wheel did not make possible the peformance of large-scale industrial work, work that would have been either impossible or marginal with the animate power sources that were its alternatives. While it was capable of being developed into a more powerful prime mover, . . . it was not. It remained largely an item of peasant culture."

7. Bennett and Elton, *Corn Milling*, II, 10–11, 16.

8. Quoted in ibid., p. 16.

9. For an explanation of why the Romans took so long to apply and diffuse the knowledge of their improved watermill, see Reynolds, *Stronger Than a Hundred Men*, pp. 32–35.

10. Ibid., pp. 36, 40–42; Usher, *Mechanical Inventions*, p. 176.

11. Bennett and Elton, *Corn Milling*, II, pp. 31–37; R. Thurston Hopkins, *Old Watermills and Windmills*, p. 13; Usher, *Mechanical Inventions*, pp. 168–69.

12. See Reynolds, *Stronger Than a Hundred Men*, p. 7.

13. Bennett and Elton, *Corn Milling*, II, pp. 79–100; Usher, *Mechanical Inventions*, pp. 179–80.

14. Reynolds, *Stronger Than a Hundred Men*, p. 51.

15. Jean Gimpel, *The Medieval Machine: The Industrial Revolution of the Middle Ages*, pp. 10, 16–17.

16. David C. Douglas and George W. Greenaway, eds., *English Historical Documents*, Vol. II, *1042–1189*, pp. 846–50. Quotes from J. H. Clapham, ibid., p. 270.

17. H. C. Darby, *Domesday England*, pp. 270–75, 361.

18. Usher, *Mechanical Inventions*, pp. 180–82, 184–86; Reynolds, *Stronger Than a Hundred Men*, p. 48.
19. Reynolds, *Stronger Than a Hundred Men*, p. 65.
20. Louis C. Hunter, *A History of Industrial Power in the United States, 1780–1930*, Vol. I, *Water Power in the Century of the Steam Engine*, p. 104.
21. Ibid., pp. 104–105; E. N. Hartley, *Ironworks on the Saugus: The Lynn and Braintree Ventures of the Company of Undertakers of the Ironworks in New England*, p. 12.
22. Hunter, *Industrial Power*, p. 105.
23. Charles Howell, "Colonial Watermills in the Wooden Age" in *America's Wooden Age: Aspects of its Early Technology*, ed. Brook Hindle, p. 307; also, Hunter, *Industrial Power*, p. 307.
24. Hunter, *Industrial Power*, pp. 305–28.
25. Ibid., p. 321.
26. Oliver Evans, *The Young Mill-Wright and Miller's Guide* 13th ed.

Water Mill Vignettes

1. Milton Rafferty, *The Ozarks: Land and Life*, pp. 3–4.
2. Louis C. Hunter, *A History of Industrial Power in the United States*, Vol. I, *Waterpower in the Century of the Steam Engine*, pp. 118–19, 123.
3. Janet C. Rowe, "The Lock Mill, Loose Creek, Missouri: The Center of a Self-Sufficient Community, 1848," *Missouri Historical Review* 75 (April, 1981): 320–21.

4. Maxine Curtis, "The Old Red Mill at Alley Spring," *Ozarks Mountaineer* 21 (July, 1973): 27; Hadley K. Irvin, "Missouri Parks Board Revives Scenic Old Red Mill," ibid. 7 (February, 1959): 11; M. E. Oliver, *Old Mills of the Ozarks: Sketches*, 2d ed., p. 21.

5. *Southeast Missourian*, July 24, 1961, pp. 1, 8; July 23, 1986, pp. 1, 4; Herman Steen, *Flour Milling in America*, pp. 126, 306–307; Don Draper, "Don Draper's Painting of Historic Old Burfordville Mill," *Ozarks Mountaineer* 6 (September, 1958): 5; Oliver, *Old Mills*, p. 24; Papers, Bollinger Mill Archives.

6. Interview with James Lalumondiere, December 4, 1986; R. N. McMullen (comp.), *An Illustrated Historical Atlas Map of Jefferson County, Mo., Carefully Compiled from Personal Examinations and Surveys*, pp. 31, 34; *Goodspeed's History of Franklin, Jefferson, Washington, Crawford and Gasconade Counties, Missouri...*, pp. 871–72.

7. Interview with Erwin Viehland, August 28, 1986; Oliver, *Old Mills*, p. 28; Norma Tynes, "Ice and Feed Mill Up for Sale after Nearly 100 Years in Family," *Meramec Journal*, November 3, 1982, pp. 8A–9A; interview with James Lalumondiere, December 4, 1986; McMullen, *Illustrated Historical Atlas Map*, p. 34.

8. Don Draper, "Don Draper's Painting of Dawt Mill that Still Grinds On," *Ozarks Mountaineer* 4 (December, 1956): 7; "Historic Dawt Mill," ibid. 18 (October, 1970): 15; Kathryn S. Love, "Old Grain Mill Turns into a Power Plant," *Saint Louis Post-Dispatch*, March 9, 1986, p. 2C; Oliver, *Old Mills*, p. 6; interview with Geredie Nesbit, September 10, 1986; interview with Tay Smith, September 10, 1986; Louise Fleming, "Ozark Streams Power Four Picturesque Mills," *Baxter Bulletin* (Mountain Home, Ark.), June 16, 1977, p. 1B.

9. Tony Czech, "Dillard Roller Mill," *Old Mill News*, 14 (Spring, 1986): 16–17; Edie Succio, "Dillard Mill Designated Historical Site," *Ozarks Mountaineer* 26 (February, 1978): 30; Kathyrn S. Love, "The Ghost That Hounds Missouri's Dillard Mill," *Saint Louis Post-Dispatch*, May 22, 1986, pp. 1F, 6F; Lennis Leonard Broadfoot, *Pioneers of the Ozarks*, p. 86; Tony Czech to author, November 3, 1986.

10. Mary L. Hahn and Blanche Reilly, *Bollinger County 1851–1976: A Commemorative History*, pp. 413–14, 433; interview with Dorothy Krieger, October 21, 1986.

11. Telephone interviews with Edna Lewis, Duard Johnson, and Linda Johnson, November 12, 1986; Lois Roper Beard, "Dryknob," in *The History of Laclede County, Missouri*, pp. 84–86; Ellen Gray Massey to author, October 31, 1986; Kirk Pearce, "Drynob—Where Time Stands Still," *Lebanon Daily Record*, September 27, 1973, p. 5; Tony Czech, "The Last Days of the Drynob Mill, Laclede County, MO," *Old Mill News* 15 (Winter, 1987): 6–7.

12. Russ How, "Side-Tripping from McCormack Lake," *Ozarks Mountaineer* 17 (May, 1969): 18–19; National Forest Service Poster, Falling Spring Mill; Missouri Division of Resources and Development, *Big Springs Country of Missouri*, Recreational Booklet No. 4 (Jefferson City: Division of Resources and Development, ca. 1940s), p. 16.

13. "Greer Springs—Another of Don Draper's Paintings," *Ozarks Mountaineer* 5 (December, 1957): 10; How, "Side-Tripping from McCormack Lake," pp. 18–19; Oliver, *Old Mills*, p. 26; William Howard Mormon, "History of Greer Mill," *Missouri Historical Review*, 66 (July, 1972): 610–21.

14. Don Blisard, "Hammond—Another Draper Painting of Water Mills," *Ozarks Mountaineer* 5 (May, 1957), 7; Oliver, *Old Mills*, p. 8; How, "Side-Tripping from McCormack Lake," pp. 18–19.

15. Dorothy J. Caldwell, *Missouri Historic Sites Catalogue*, p. 119; Clay M. Anderson, "Hodgson Mill on Bryant Creek," *Ozarks Mountaineer* 31 (June, 1983): 36–37; "Old Watermills Give Romance to Ozark County, Missouri," *Ozarks Mountaineer* 12 (May, 1965): 13; Oliver, *Old Mills*, index; Roscoe Misselhorn, *Misselhorn's Pencil Sketches of Missouri*, ed. Harry M. Hagen, pp. 6–7; Missouri Division of Resources and Development, *White River Country of Missouri*, Recreational Booklet No. 1 (Jefferson City: Division of Resources and Development, ca. 1940s), p. 10; Fleming, "Ozark Streams Power Four Picturesque Mills," p. 1B.

16. Richard Garrity, "Halloween at a Gristmill," *Ozarks Mountaineer* 22 (October, 1974): 18; Otis Hays, Jr., "The Puzzlements of Jolly Mill," ibid. 29 (August, 1981): 46–48; Oliver, *Old Mills*, p. 19; Juanita Hays and Otis Hays, "Isbell's Distillery and Jolly Mill," *Old Mill News* 8 (Fall, 1985), 6–7; interview with L. Orville Goodman, November 20, 1986; Nellie Alice Mills, *Historic Spots in Old Barry County*, pp. 11–13, 17–18.

17. Oliver, *Old Mills*, index; Otis Dunn, "Old McDowell Grist Mill—Another of Don Draper's Paintings," *Ozarks Mountaineer* 5 (July, 1957): p. 9; Mills, *Historic Spots*, pp. 45–46.

18. Malcolm C. Drummond, *Historic Sites in Franklin County*, p. 92; Henry Gottlieb Kiel, *Centennial Biographical Directory of Franklin County*, p. 207; Helen Vogt to author, November 3, 1986; Missouri Division of Resources and

Development, *Meramec Valley of Missouri*, Recreational Booklet (Jefferson City: Division of Resources and Development, ca. 1940s), p. 15.

19. Charles Sherman, "Old Appleton: A Quiet Missouri Town with the Charm of Unhurried Age," *Saint Louis Post-Dispatch*, July 17, 1966, "Pictures," pp. 1–7; Jerry May, "One Man's Dream: Restoring Old Appleton's Mill," *Southeast Missourian*, April 11, 1967, pp. 1, 5; Delecia B. Huitt, "Down by the Old Mill Stream: A History of the Old Appleton Mill, unpublished manuscript, Southeast Missouri State University; Oliver, *Old Mills*, pp. 30–31; Caldwell, *Missouri Historic Sites Catalogue*, p. 23; Misselhorn, *Misselhorn's Pencil Sketches of Missouri*, pp. 16–17; interview with R. C. ("Rip") Schnurbusch, September 18, 1986; Tom Neumeyer, "Old Appleton," *Cash-Book Journal* (Jackson, Mo.), July 23, 1975, p. 5; K. J. H. Cochran, "Old Appleton Mill and Bridge," *Tipoff*, 15 (January, 1984): 16–18.

20. Interview with Gorton Thomas, November 20, 1986; interview with Manford Troxel, November 20, 1986; Mills, *Historic Spots*, pp. 15–17.

21. Bonnie Howlett Bilisoly, "Visiting the Past at Schlicht's Mill," *Ozarks Mountaineer* 24 (July, 1976): 24–25; Tom Chesser, "Schlicht's Mill: A Monument to Another Era," ibid. 26 (July, 1978): 26–27; Kathryn S. Love, "Down by the Old Mill Stream," *Saint Louis Post-Dispatch*, May 22, 1986, pp. 1F, 6F; Patsy Watts, "Schlicht Spring and Mill," *Bittersweet*, 5 (Summer, 1978): 26–31; Mabel Manes Mottaz, *Lest We Forget: A History of Pulaski County, Missouri, and Fort Leonard Wood*, pp. 24, 69–70.

22. Clay M. Anderson, "The Ozark Water Mill Trail," *Ozarks Mountaineer* 17 (May, 1969): 16–17; Ruth Bowler, "Unusual Names in the Ozarks: Topaz," ibid.

15 (October, 1967): 7; Don Draper, "Topaz Mill of Yesteryear—Another Don Draper Painting," ibid. 6 (July, 1958): 11; Oliver R. Griffin, "A Trip to Topaz Mill," ibid. 22 (December, 1974): 36; Oliver, *Old Mills*, p. 23; interview with Joe W. O'Neal, owner of Topaz Mill, September 10, 1986.

23. Missouri Division of Resources and Development, *White River Country of Missouri*, Recreational Booklet No. 1 (Jefferson City: Division of Resources and Development, ca. 1940s), p. 16; Oliver, *Old Mills*, p. 9 and index; Caldwell, *Missouri Historic Sites Catalogue*, p. 120; Fleming, "Ozark Streams Power Four Picturesque Mills," p. 1B.

Epilogue: Going to Mill

1. R. N. McMullen (comp.), *An Illustrated Historical Atlas Map of Jefferson County, Mo., Carefully Compiled from Personal Examinations and Surveys*, pp. 31, 34.

2. Tony Czech, "Dillard Roller Mill," *Old Mill News* 14 (Spring, 1986): 16–17; Jack Smoot, "The Effects of Milling Technology Improvements on the 19th Century Consumer" (unpublished manuscript, 1985), p. 10.

3. Quoted in Louis C. Hunter, *A History of Industrial Power in the United States, 1780–1930*, Vol. I., *Waterpower in the Century of the Steam Engine*, pp. 12, 13.

4. Quoted in Lennis Leonard Broadfoot, *Pioneers of the Ozarks*, p. 88.

5. Janet C. Rowe, "The Lock Mill, Loose Creek, Missouri: The Center of a Self-Sufficient Community, 1848–1890," *Missouri Historical Review* 75 (April, 1981): 320–21.

6. Interview with Manford Troxel, November 20, 1986.
7. Kirk Pearce, "Drynob—Where Time Stands Still," *Lebanon Daily Record*, September 27, 1973, p. 5.
8. Interview with Gorton Thomas, November 20, 1986.
9. Nellie Alice Mills, *Historic Spots in Old Barry County*, pp. 16–17.
10. Interview with Gorton Thomas, November 20, 1986.

Glossary

Bed Stone. The fixed, bottom millstone against which the revolving runner stone grinds grain.
Bolting. The process of sifting flour into different degrees of fineness.
Breast Wheel. A vertical waterwheel turned by running water striking the buckets of the wheel either immediately above or below the level of its axle. If above, the wheel is called a high breast wheel; if below, a low breast wheel.
Buckets. Water receptacles around the rim of the wheel. The impact and weight of the water on the receptables drive the wheel.
Cogwheel. A wheel with sturdy cogs or teeth inserted at intervals into its interior rim (the side away from the waterwheel) parallel to the shaft of the waterwheel. The wheel is fixed to the shaft.
Custom Mill. A small water mill of limited capacity that grinds on demand for customers in its immediate area, for which they pay a toll in kind.
Eye. The hole in the center of the runner stone.
Face Wheel. *See* cogwheel.
Finest. Description of a miller's best grade of flour.
Flume. A trough of wood or metal used to direct water to a waterwheel.

Forebay. The part of a millpond or millrace from which water is immediately released to run a waterwheel or turbine.

French Buhrstone. A siliceous rock or stone from which the best, most prized millstones were made.

Furrows. The intricate pattern of cuts on a millstone that shears off the outer husk of grain, moves the grain onto the "land," directs the ground product to the outer edge of the millstone, and reduces the heat of grinding by allowing air to the face of the stone.

Greek Mill. An early and simple form of a water mill. It consists of a vertical shaft topped by a runner stone turned by running water striking blades or paddles attached to the bottom of the shaft.

Grist. Grain (all types) taken to a mill for grinding.

Gristmill. A building with machinery for grinding grain into meal or flour.

Head. A mass of water in motion, as in a stream, used to power a waterwheel or turbine; or, the downward distance water falls to turn a wheel or turbine.

High Breast Wheel. *See* breast wheel.

Hopper. A funnel-shaped wooden box used to direct grain to the eye of the runner stone.

Horizontal Mill. *See* Greek mill.

Land. Crushing or grinding surface of a millstone located between the furrows.

Lantern Gear. A gear pinion, resembling a lantern, that has cylindrical bars instead of teeth inserted between two parallel disks. It is used to redirect the flow of power from a cogwheel in millworks.

Low Breast Wheel. *See* breast wheel.

Merchant Mill. A mill with the capacity to produce meal and flour in substantial quantities for sale to outlying districts and distant markets.
Middlings. A middle- or medium-quality flour; or, a by-product of flour milling composed of several grades of granular particles used as animal feed.
Millrace. A canal in which water flows both to and from a waterwheel; or, generally the current that powers the wheel.
Millrun. *See* flume.
Mortar and Pestle. A primitive handmill consisting of a small bowl-shaped vessel, the mortar, usually made of stone or wood, and a club-shaped device, the pestle, used for pounding and crushing grain within the mortar.
Mortise-joint Construction. Method used in building older mill structures that employs joints made by a groove (mortice) in one part and a tongue (tenon) in the other, the latter being inserted into the mortice and locked with a wooden pin or treenail.
Norse Mill. *See* Greek mill.
Overshot Wheel. A vertical waterwheel powered by the impact and weight of running water falling from above the wheel and striking buckets positioned on the rim of the wheel. Viewed with the water flowing left to right, the standard overshot wheel will turn clockwise. *See also* pitchback wheel.
Penstock. A trough or passageway of wood or metal which directs water to a turbine.
Pit Saw. A long handsaw with handles at each end used for cutting logs lengthwise into slabs or planks. One man worked the saw above the log; another worked below the log while standing in a pit.
Pitchback Wheel. An overshot wheel in which the impact and weight of the

water on the wheel is at a point short of the axis of rotation of the wheel, forcing it to revolve counterclockwise, as viewed with the stream flowing from left to right.

Plumping Mill. A crude water mill used for pounding grain in a mortar. It consists of a pounder, or pestle, and a bucket on opposite ends of a pole that rested on a fulcrum. When filled with water the bucket lifted the pounder; the pounder fell when the bucket was emptied.

Rafting Chute. A chute built into a milldam that allows rafts of logs or railroad ties to float down the millstream, thereby permitting another form of commercial use of the stream.

Rollers. A modern replacement for the millstones which improved the efficiency of water mills and the quality of products they produced.

Rotary Quern Mill. A primitive mill consisting of a bedstone and a runner stone with a handle attached. The stones are usually enclosed.

Shorts. Lowest grade of flour.

Sluice Gate. A gate which controls the amount of water directed against the wheel or turbine.

Sluiceway. *See* flume.

Stone Quern. A primitive, hand-operated mill consisting of a fixed stone and a runner stone that is rotated.

Stump Mill. A primitive mill consisting of a pounder attached to a sapling, the spring action of which is used to help lift the pounder for crushing grain placed in a hollow stump.

Timber Cribbing. A boxlike structure of logs, usually filled with stones or rubble, used in the construction of milldams.

Toll. Payment in kind, usually collected as a percentage of the meal or flour, to the miller for grinding.

Tub Wheel. A horizontal waterwheel resembling both the Greek mill and the modern turbine. It is enclosed in a reinforced, half-barrellike structure into which water flows from a sluiceway.

Turbine. An advanced tub wheel that generates more power with greater efficiency than a waterwheel. It is a waterpowered rotary engine generally made with a series of curved vanes on a shaft enclosed within a casing. In the casing are redirecting vanes and passageways that permit the intake and release of water in the desired manner.

Undershot Wheel. A vertical wheel turned counterclockwise by water striking paddles or buckets at a point below the axle of the wheel.

Wallower. *See* lantern gear.

Selective Bibliography

Books

Arthur, George Clinton. *Backwoodsmen, Daring Men of the Ozarks, Including the Life of Nathaniel (Stub) Borders*. Boston: The Christopher Publishing House, 1940.

Baer, Howard, et al. *Missouri, Heart of the Nation: A Pictorial Record by Fourteen American Artists*. New York: American Artists Group, 1947.

Bathe, Greville, and Dorothy Bathe. *Oliver Evans: A Chronicle of Early American Engineering*. Philadelphia: The Historical Society of Pennsylvania, 1935.

Beard, Lois Roper, et al. (comps.). *The History of Laclede County, Missouri*. Vol. I. Tulsa, Okla.: Heritage Publishing Co., 1979.

Bennett, Richard, and John Elton. *History of Corn Milling*. 4 vols. 1898–1904; reprint New York: Burt Franklin, [1964].

Bowden, Witt. *Industrial Society in England Towards the End of the Eighteenth Century*. 2d ed. London: Frank Cass and Co., 1965.

Bowers, Douglas. *A List of References for the History of Agriculture in the United States, 1790–1840*. Davis, Calif.: Agricultural History Center, 1969.

Breuer, James Ira. *Crawford County and Cuba, Missouri, with a Supplement: Oak Grove School, 1858–1945*. Cape Girardeau, Mo.: Ramfre Press, 1972.

Broadfoot, Lennis Leonard. *Pioneers of the Ozarks*. Caldwell, Idaho: The Caxton Printers, 1944.

Bryan, William Smith, and Robert Rose. *A History of the Pioneer Families of Missouri, with Numerous Sketches, Anecdotes, Adventures, etc., Relating to Early Days in Missouri*. Saint Louis: Bryan, Brand and Co., 1876.

Bryant, William Cullen (ed.). *Picturesque America; or, The Land We Live In*. 2 vols. New York: D. Appleton and Co., 1872.

Caldwell, Dorothy J., ed. *Missouri's Historic Sites Catalogue*. Columbia, Mo.: State Historical Society of Missouri, 1963.

Clapham, J. H. *A Concise Economic History of Britain from the Earliest Times to 1750*. Cambridge: Cambridge University Press, 1949.

Collingwood, R. G., and J. N. L. Myres. *Roman Britain and the English Settlements*. Vol. I. *The Oxford History of England*. London: Oxford University Press, 1963.

Crockett, Norman L. *The Woolen Industry of the Midwest*. Lexington: University of Kentucky Press, 1970.

Darby, H. C. *Domesday England*. London: Cambridge University Press, 1977.

Daumas, Maurice (ed.). *A History of Technology and Invention*. Trans. Eileen B. Bennessy. 2 vols. New York: Crown Publishers, 1969.

Douglas, David C., and George W. Greenaway (eds.). *English Historical Documents*. Vol. II. *1042–1189*. London: Eyre & Spottiswoode, 1953; reprint, New York: Oxford University Press, 1968.

Drummond, Malcolm C. *Historic Sites in Franklin County*. N.p.: Harland Bartholomew and Associates, 1978.

Eckberg, Carl. *Colonial Ste. Genevieve: An Adventure on the Mississippi Frontier*. Gerald, Mo.: The Patrice Press, 1985.

Evans, Oliver. *The Young Mill-Wright and Millers Guide*. Philadelphia: Carey, Lea & Blanchard, 1834; reprint, New York: Arno Press, 1972.

Fairbairn, William. *Treatise on Mills and Millworks*. 2 ed. 2 vols. London: Longman, Green, Longman, Roberts and Green, 1864.

Ferguson, Eugene S. *Oliver Evans: Inventive Genius of the American Industrial Revolution*. Greenville, Del: Hagley Museum and the University of Delaware, 1980.

Firm, R. William. *An Introduction to Domesday Book*. London: Longmans, Green and Co., 1963.

———. *The Domesday Inquest and the Making of Domesday Book*. London: Longmans and Co., 1963.

Fox, William; Bill Brooks; and Janice Tyrwhitt. *The Mill*. Boston: New York Graphic Society, 1976.

Gates, Paul W. *The Farmers' Age: Agriculture, 1815–1860*. The Economic History of the United States Series, Vol. 3. New York: Holt, Rinehart, and Winston, 1960.

Goodspeed's History of the Ozark Region; Comprising a Condensed General History, a Brief Descriptive History of Each County, and Numerous Biographical Sketches of Prominent Citizens of Such Counties. Chicago: Goodspeed's Publishing Co., 1894; reprint, Cape Girardeau, Mo.: Ramfre Press, 1956.

Gimpel, Jean. *The Medieval Machine: The Industrial Revolution of the Middle Ages*. New York: Holt, Rinehart and Winston, 1976.

Hahn, Mary L., and Blanche Reilly. *Bollinger County, 1851–1976: A Commemorative History*. Marceline, Mo.: Wolsworth Publishing Co., for the Bollinger County Bicentennial Commission, 1977.

Hartley, E. N. *Ironworks on the Saugus: The Lynn and Braintree Ventures of the Company of Undertakers of the Ironworks in New England*. Norman: University of Oklahoma Press, 1957.

Hindle, Brook (ed.). *America's Wooden Age: Aspects of Its Early Technology*. Tarrytown, N.Y.: Sleepy Hollow Restorations, 1976.

Historic Missouri: A Pictorial Narrative. Columbia, Mo.: State Historical Society of Missouri, 1977.

History of Franklin, Jefferson, Washington, Crawford and Gasconade Counties, Missouri, from the Earliest Times to the Present; Together with Sundry Personal, Business and Professional Sketches and Numerous Family Records; besides a Valuable Fund of Notes, Original Observations, etc. Chicago: Goodspeed Publishing Company, 1888.

Hopkins, R. Thurston. *Old Watermills and Windmills*. London: Philip Allan and Co., n.d.

Houck, Louis. *A History of Missouri from the Earliest Explorations and Settlements until the Admission of the State into the Union*. 3 vols. Chicago: R. R. Donnelley and Sons, 1908.

———. *Memorial Sketches of Pioneers and Early Settlers of Southeast Missouri*. Cape Girardeau, Mo.: n.p., 1915.

Howell, Charles, and Alan Keller. *The Mill at Philipsburg Manor, Upper Mills and a Brief History of Milling*. Tarrytown, N.Y.: Sleepy Hollow Restorations, 1977.

Hunter, Louis C. *A History of Industrial Power in the United States, 1780–1930*. 2 vols. Charlottesville: University Press of Virginia, for the Eleutherian Mills–Hagley Foundation, 1979, 1987.

Kiel, Henry Gottlieb. *Centennial Biographical Directory of Franklin County*. N.p.: n.p., 1925.

Kulhmann, Charles B. *The Development of the Flour-Milling Industry in the United States, with Special Reference to the Industry in Minneapolis*. Boston: Houghton-Mifflin, 1929.

Lyle, Wes, and John Hall. *Missouri: Faces and Places*. Lawrence: Regents Press of Kansas, 1977.

Magill, Arthur Clay. *Geography and Geology of the Southeast Missouri Lowlands*. Ed. Felix Eugene Snider. Cape Girardeau, Mo.: Ramfre Press, 1958.

Massey, Elley Gray. *Bittersweet Country*. Garden City, N.Y.: Anchor Press/Doubleday, 1978.

Matthews, Norval M. *The Promised Land: A Story about Ozark Mountains and the Early Settlers of Southwest Missouri*. Point Lookout, Mo.: School of the Ozarks Press, 1974.

McMullen, R. N. (comp.). *An Illustrated Historical Atlas Map of Jefferson County, Mo., Carefully Compiled from Personal Examinations and Surveys*. N.p.: Brink, McDonough and Company, 1876.

Mills, Nellie Alice. *Historic Spots in Old Barry County*. Monett, Mo.: The Free Will Baptist Gem, 1952.

Missilhorn, Roscoe. *Misselhorn's Pencil Sketches of Missouri*. Ed. Harry M. Hagen. Saint Louis: Riverside Press, 1974.
Mottaz, Mabel Manes. *Lest We Forget: A History of Pulaski County, Missouri, and Fort Leonard Wood*. Springfield, Mo.: Cain Printing Company, 1960.
Nettles, Curtis P. *The Roots of American Civilization: A History of American Colonial Life*. Crofts American History Series. New York: Appleton-Century-Crofts, Inc., 1938.
North, Douglas C. *The Economic Growth of the United States, 1790–1860*. Englewood Cliffs, N.J.: Prentice Hall, 1961.
Oliver, John W. *History of American Technology*. New York: The Ronald Press Co., 1956.
Oliver, M. E. *Old Mills of the Ozarks: Sketches*. 2d ed. N.p., 1971.
Parrish, William E.; Charles T. Jones; and Lawrence O. Christensen. *Missouri: The Heart of the Nation*. Saint Louis: Forum Press, 1980.
Rafferty, Milton D. *Missouri: A Geography*. Boulder, Colo.: Westview Press, 1983.
———. *The Ozarks: Land and Life*. Norman: University of Oklahoma Press, 1980.
Reynolds. John. *Windmills and Watermills*. New York: Praeger, 1975.
Reynolds, Terry S. *Stronger than a Hundred Men: A History of the Vertical Water Wheel*. Baltimore: The Johns Hopkins University Press, 1983.
Russell, Jesse Lewis. *Behind these Ozark Hills: History, Reminiscences, Traditions Featuring the Authors's Family. . . . Biographical Sketches of Outstanding Descendants of Pioneers*. New York: The Hobson Book Press, 1947.

Sass, Jon A. *The Versatile Millstone: Workhorse of Many Industries*. Knoxville, Tenn.: Society for the Preservation of Old Mills, 1984.
Sauer, Carl O. *The Geography of the Ozark Highland of Missouri*. N.p., n.d.
Schultz, Gerard. *Early History of the Northern Ozarks*. Jefferson City, Mo.: Midland Printing Company, 1937.
Sechler, Earl Truman. *Leaves from an Ozark Journal*. 3 vols. Springfield, Mo.: Westpoint Press, 1969–1971.
Selby, Paul O. *A Bibliography of Missouri County Histories and Atlases*. 2d ed. Kirksville, Mo.: Northeast Missouri State Teachers College, 1966.
Singer, Charles J., et al. (eds.). *A History of Technology*. 5 vols. Oxford: Clarendon Press, 1954–1978.
Steen, Herman. *Flour Milling in America*. Westport, Conn.: Greenwood Press, 1963.
Storck, John and Walter D. Teague. *Flour for Man's Bread*. Minneapolis: University of Minnesota Press, 1952.
Thomas, George Brown. *True Tales of the Ozarks: Memories of a Southern Missouri Boyhood after the Civil War*. New York: Exposition Press, 1963.
Usher, Abbott P. *A History of Mechanical Inventions*. Rev. ed. Cambridge: Harvard University Press, 1954.
———. *The Industrial History of England*. Boston: Houghton Mifflin Co., 1920.
U.S., House. *Report on the Manufactures of the United States*. Misc. Doc. 42, Part 2. 47th Cong., 2d sess., 1883.
———. *Reports on the Water-Power of the United States*. Misc. Doc. 42, Parts 16 and 17. 47th Cong., 2d sess., 1885 and 1887.

WPA Writers' Project. *Missouri: A Guide to the "Show Me" State*. New York: Duell, Sloan and Pearce, 1941.

White, Jr., Lynn. *Medieval Technology and Social Change*. Oxford: Clarendon Press, 1962.

Zimiles, Martha Murray. *Early American Mills*. New York: Clarkson N. Potter, Inc., 1973.

Articles

Anderson, Clay M. "Hodgson Mill on Bryant Creek." *Ozarks Mountaineer* 31 (June, 1983): 36–37.

———. "The Ozark Water Mill Trail." *Ozarks Mountaineer* 17 (May, 1969): 16–17.

Anderson, Russell H. "The Technical Ancestry of Grain Milling Devices." *Agricultural History* 12 (July, 1938): 256–70. Reprint from *Mechanical Engineering* 57 (1935): 611–17.

———. "Advancing Across the Eastern Mississippi Valley." *Agricultural History* 17 (April, 1943): 97–104.

Bell, Ovid. "Pioneer Life in Calloway County." *Missouri Historical Review* 21 (January, 1927): 156–65.

Bilisoly, Bonnie Howlett. "Visiting the Past at Schlicht's Mill." *Ozarks Mountaineer* 24 (July, 1976): 24–25.

Blissard, Don. "Hammond—Another Draper Painting of Water Mills." *Ozarks Mountaineer* 5 (May, 1957): 7.

Bowler, Ruth. "Unusual Names in the Ozarks: Topaz." *Ozarks Mountaineer* 15 (October, 1967): 7.

Brinkman, Grover. "A Soul Soaking Sight." *Ozarks Mountaineer* 24 (May, 1976): 17.

Caldwell, Dorothy J. "Missouri's National Historic Landmarks: Watkins Mill." Part III. *Missouri Historical Review* 63 (April, 1969): 364–77.

Chesser, Tom. "Schlicht's Mill: A Monument to Another Era." *Ozarks Mountaineer* 26 (June, 1978): 26–27.

Crockett, Norman L. "The Marketing of Wool in the Nineteenth Century: The Case of the Middle West." *Agricultural History* 42 (October, 1968): 315–26.

Curtis, Maxine. "The Old Red Mill at Alley Spring." *Ozarks Mountaineer* 21 (July, 1973): 27.

Czech, Tony. "Dillard Roller Mill." *Old Mill News* 14 (Spring, 1986): 16–17.

———. "The Last Days of the Drynob Mill, Laclede County, MO," *Old Mill News* 15 (Winter, 1987): 6–7.

Dick, Everett. "Going Beyond the Ninety-Fifth Meridian." *Agricultural History* 17 (April, 1953): 105–11.

"Don Draper's Painting of Historic Old Burfordville Mill." *Ozarks Mountaineer* 6 (September, 1958): 5.

Draper, Don. "Draper's Painting of Dawt Mill that Still Grinds On." *Ozarks Mountaineer* 4 (December, 1956): 7.

———. "Falling Springs Mill—Another of Don Draper's Paintings." *Ozarks Mountaineer* 5 (November, 1957): 6.

———. "Topaz Mill of Yesteryear—Another Don Draper Painting." *Ozarks Mountaineer* 6 (July, 1958): 11.

Dunn, Otis. "Old McDowell Grist Mill—Another of Don Draper's Paintings." *Ozarks Mountaineer* 5 (July, 1957): 9.

Evans, Priscilla Ann. "Merchant Gristmills and Communities, 1820–1880: An Economic Relationship." *Missouri Historical Review* 68 (April, 1974): 317–26.

Fleming, Louise. "Ozark Streams Power Four Picturesque Mills." *Baxter Bulletin* (Mountain Home, Ark.), June 16, 1977, p. 1B.

Fitch, Jim. "Ozark County Offers the Real Ozarks." *Ozarks Mountaineer* 30 (April, 1982): 60–61.

Garrity, Richard. "Halloween at a Gristmill." *Ozarks Mountaineer* 22 (October, 1974): 18.

"Greer Springs—Another of Don Draper's Mill Paintings." *Ozarks Mountaineer* 5 (December, 1957): 10.

Griffin, Oliver R. "A Trip to Topaz Mill." *Ozarks Mountaineer* 22 (December, 1974): 36.

Gripka, Phyllis Harrell. "A Man and a Mill." *Ozarks Mountaineer* 28 (April, 1980): 26.

Hasse, Larry. "Watermills in the South: Rural Institutions Working Against Modernism." *Agricultural History* 58 (July, 1984): 280–95.

Hays, Jr., Otis. "The Puzzlements of Jolly Mill." *Ozarks Mountaineer* 29 (August, 1981): 46–48.

Hays, Juanita, and Otis Hays. "Isbell's Distillery and Jolly Mill." *Old Mill News* 8 (Fall, 1985): 6–7.

"Historic Dawt Mill." *Ozarks Mountaineer* 18 (October, 1970): 15.

How, Russ. "Side-Tripping from McCormack Lake." *Ozarks Mountaineer* 17 (May, 1969): 18–19.

Irwin, Hadley K. "Missouri Park Board Revives Scenic Old Red Mill." *Ozarks Mountaineer* 7 (February, 1959): 11.

Loehr, Rodney. "Moving Back from the Atlantic Seaboard." *Agricultural History* 17 (April, 1943): 90–96.

Love, Kathyrn. "Old Grain Mill Turns into a Power Plant." *Saint Louis Post-Dispatch*, March 9, 1986, p. 2C.

———. "The Ghost that Hounds Missouri's Dillard Mill." *Saint Louis Post-Dispatch*, May 22, 1986, pp. 1F, 6F.

May, Jerry. "One Man's Dream: Restoring Old Appleton's Mill." *Southeast Missourian*, April 11, 1967, pp. 1, 5.

Neumeyer, Tom. "Old Appleton." *Cash-Book Journal* (Jackson, Mo.), July 23, 1975, p. 5.

McRaven, Charles. "Ozark County Mills Revisited." *Ozarks Mountaineer* 22 (September, 1974): 24–25, 32.

Mormon, William Howard. "History of the Greer Mill." *Missouri Historical Review* 66 (July, 1972): 610–21,

"Old Watermills Give Romance to Ozark County, Missouri." *Ozarks Mountaineer* 12 (May, 1965): 13.

"Ozark Water Mill Trail: Mo. 95—Mountain Grove to Theodosia." *Ozarks Mountaineer* 15 (October, 1967): 28–29.

Pearce, Kirk. "Drynob—Where Time Stands Still," *Lebanon Daily Record*, September 27, 1973, p. 5.
Ross, Earle D. "Retardation in Farm Technology Before the Power Age." *Agricultural History* 30 (January, 1956): 11–18.
Rowe, Janet C. "The Lock Mill, Loose Creek, Missouri: The Center of Self-Sufficient Community, 1848–1890." *Missouri Historical Review* 75 (April, 1981): 285–93.
Sharrer, G. Terry. "The Merchant-Miller: Baltimore's Flour-Milling Industry, 1783–1860." *Agricultural History* 56 (January, 1982): 138–50.
Sherman, Charles. "Old Appleton: A Quiet Town with the Charm of Unhurried Age." *Saint Louis Post-Dispatch*, July 17, 1966, "Pictures," pp. 1–7.
Stovsky, Renee. "Down by the Old Mill Stream." *Saint Louis Globe-Democrat*, November 3, 1978, p. 3D.
Succio, Edie. "Dillard Mill Designated Historical Site." *Ozarks Mountaineer* 26 (February, 1978): 30.
Suggs, Jr., George G. "Watermills: The Paintings of Jake Wells." *Missouri Life* 6 (July–August, 1978): 30–37.
Tynes, Norma. "Ice and Feed Mill Up for Sale after Nearly 100 Years in Family." *Meramec Journal*, November 3, 1982, pp. 8A–9A.
"Views from the Past: Missouri Mills." *Missouri Historical Review* 63 (October, 1968): 90–91.
Voss, Stuart F. "Town Growth in Central Missouri, 1815–1880: An Urban Chaparral." Part I. *Missouri Historical Review* 64 (October, 1969): 64–80.

———. "Town Growth in Central Missouri." Part II. *Missouri Historical Review* 64 (January, 1970): 197–217.
———. "Town Growth in Central Missouri." Part III. *Missouri Historical Review* 64 (April, 1970): 322–50.
Watts, Patsy. "Schlicht Spring and Mill." *Bittersweet* 5 (Summer, 1978): 26–31.
Williamson, Hugh P., and Arnold Bedsworth. "Duley's Mill." *The Bulletin* 21 (April, 1965): 245–48.

Interviews

Rene Dellamano, September 18, 1986
L. Orville Goodman, November 20, 1986
Duard Johnson (telephone), November 12, 1986
Linda Johnson (telephone), November 12, 1986
Dorothy Krieger, October 21, 1986
James Lalumondiere, December 4, 1986
Edna Lewis (telephone), November 12, 1986
Joe W. O'Neal, September 10, 1986
Geredie Nesbit, September 10, 1986
R. C. ("Rip") Schnurbusch, September 18, 1986
Tay Smith, September 10, 1986
Gorton Thomas, November 20, 1986
Manford Troxel, November 20, 1986
Erwin Viehland, August 28, 1986

Index

Aid, Charles T.: 102, 103
Aid-Hodgson Mill: *see* Hodgson Mill
Alley, John: 45
Alley Spring: capacity of, 44; enterprises around, 44; name change of, 46
Alley Spring Mill: 46, 99, 164; origin of, 43; location of, 43; early character of, 43; changing ownership of, 43–45; innovation on, by George McCaskill, 44; as an agent of community, 46; present status of, 162
Alsop, James N.: 50
Alton, Mo.: 86
Anglo-Saxon England: 28
Apple Creek: 63, 122; waterpower potential of, 116
Apple Creek, Mo.: origin of, 118
Appleton, Mo.: 118
Appleton City, Mo.: 118

Arkansas: 39
Arkansas River: 39
Atlantic and Pacific Railroad: 127
Aube, France: 28

Baker (McManus) Mill: 10
Barker, Noble: 136
Barksdale, Ike: 43
Barksdale Spring: name change of, to Alley Spring, 45–46
Barry County, Mo.: 104, 108, 160
Baxter Springs, Kans.: 106
Beehler, Conrad: 60
Bennett, Richard: 26
Big River: 53, 59, 60, 62, 63, 146
Black River: 8, 39
Bollinger, George Frederick: 47, 51, 74; as entrepreneur, 145–46

Bollinger, Mathias: 74; engineering skill of, 76; as entrepreneur, 145–46
Bollinger, Moses: 76
Bollinger County, Mo.: 76, 78
Bollinger-Dolle Mill: *see* Dolle Mill
Bollinger Mill: 10, 12, 37, 40, 51, 74, 79, 86, 131, 138, 145, 155, 164; origin of, 47, 49; changing ownership of, 49–50, 51; evolution of, 49–50; improvement in, 50; merger of, to Cape County Milling Company, 50; location of, 51; as example of mill transformation, 51; hazards to, 52; present status of, 162
Bollinger Mill State Historic Site, formation of: 51
Bourbeuse River: 63, 112
Britain: 27, 28
British Isles: 26
Broadfoot, Lennis L.: 69–70
Brown, George I.: 107
Brown, J. W.: 89
Brown, Walter: 89
Bryant Creek: 101
Buchheit, Joseph: 119, 120
Bunce, O. B.: 4, 6
Burford, Solomon R.: 49
Burfordville, Mo.: 47, 74

Burfordville Mill: *see* Bollinger Mill
Byrne, M. F., description of, as typical mill entrepreneur: 54, 56
Byrne, Patrick: 54
Byrne Mill: 59
Byrnesville, Mo.: 54; growth of, 56; as a custom mill, 56; improvement to, 56; decline of, 56–57; new use of, 57; location of, 57; present status of, 161
Byrnesville Mill: 59, 146, 155, 156, 164; origin of, 53–54; improvement on, 54

Cabool, Mo.: 132
Caldwell, John: 64
Cape County Historical Society: 51
Cape County Milling Company: 50, 51
Cape Girardeau, Mo.: 144
Cape Girardeau County, Mo.: 51, 116, 118
Capps Creek: 104, 123
Carter County, Mo.: 134
Casey, Tom: 158
Cattle mill: *see* mills, early examples of
Cedar Grove Mill: 150
Cedar Hill, Mo.: 59, 62, 156
Cedar Hill Ice and Feed Mill: 62
Cedar Hill Mill: 63, 155, 164; location of, 59; origin of, 60; description of,

61; change of power sources in, 61; production changes in, 61; ownership changes in, 62; present status of, 162–63
Christ: 16
Civil War: 53, 69, 78, 90, 92, 101, 106, 125, 130, 138, 152; effect of, on water mills, 82
Cody, John: 138
Cottrell, Joseph Dillard: 69
Crawford County, Mo.: 72
Crocker, Mo.: 128

Daugherty, Frank: 49
Daugherty, Sarah Bollinger: 49
Dawt Mill: 12, 86, 102, 138, 145, 155, 164; location of, 63; origin of, 63; history of, before 1900, 64; promoter of community, 66; produces electricity, 68; present status of, 162–63
Dellamano, Art: 122
Dellamano, Rene: 122
Dennig, Louis E.: 93
Dillard, Mo.: 69, 72
Dillard Mill: 131, 146, 164; origin of, 69–70; location of, 72; present status of, 162

Dillard Mill State Historic Site: 72
Dolle, John Herman: 76, 78
Dolle Mill: 10, 90, 145, 164; site selection for, 74; repossession of, by Bollinger family, 76; saving of, from destruction, 78; location of, 78; as a social center, 78; present status of, 162–63
Domesday Book, contents of, concerning water mills: 28–29
Dougherty, John: 43–45
Douglas County, Mo.: 132, 134
"Down by the Old Mill Stream," past importance of: 6–7
Drey, Leo A.: 93
Drynob, Mo.: 85; as a water mill town, 156, 158
Drynob Mill: 12, 112, 148, 153, 155, 156, 164; location of, 80; origin of, 80, 82; improvements on, 82, 84; decline of, 85; collapse of, 158

Eleven Point River: 86, 92, 93
Elton, John: 26
Eminence, Mo.: 43
England, developments in, encouraging use of Roman mill: 27, 29
English Channel: 28

Europe: 24, 29, 32, 34
Evans, Oliver, water mill innovations of: 37

Falling Spring Cemetery: 86
Falling Spring Mill: 40, 99, 101, 138, 164; location of, 86; description of, 86, 88; as a typical water mill, 89; present status of, 163–64
Flat Creek: 108, 110
France: 27, 28
Frankfurt, Ger.: 130
Franklin County, Mo.: 112
Freeman, Alabath: 132
Free Will Baptists: 158
Friend Mill: 64
Frisco Railroad: 131

Gainesville, Mo.: 99
Gasconade Mill: *see* Schlicht Mill
Gasconade River: 128
"Going to mill," importance of: 152–53
Granby, Mo.: 123
Grand River: 39
Grant, Ulysses S.: 64
Great Britain: 24
Great Depression: 70, 146

Greece: 24
Greek mill: description of, 16, 18; spread of, 18
Greer, John: 90
Greer, Samuel W.: 37, 90; as an entrepreneur, 145–46
Greer Mill: 37, 74, 99, 148, 164; origin of, 90; unusual millsite of, 92; cable system of, 92–93; changing ownership of, 93; location of, 93; present status of, 163–64
Greer Spring: 90, 92
Grudier, John W.: 94, 96

Hammersmith Ironworks, description of: 31–32
Hammond, Mo.: 96, 98; founding of, 94
Hammond Mill: 12, 155, 156, 164; location of, 94, 96; decline of, 96; description of, 96, 98; present status of, 163–64
Haskins, W. F.: 107
Hensley, John: 128
Hesse, Charles: 119
History of Corn Milling, on origins of water mills in Britain: 26
Hodgson, Alva: 63, 64, 101–103; as an entrepreneur, 145–46

Hodgson, George: 102
Hodgson Mill: 63, 99, 137, 145, 152, 164; location of, 101; origin of, 101; construction of, 102; beautiful setting of, 103; present use of, 103; present status of, 162–63
Hodgson Spring, capacity of: 102
Holeman, William: 101, 103
Horne, Phil: 68
Houck, Louis: 93
House Springs, Mo.: 59
Howell, Charles, quoted on tub mills: 32, 34
Hunter, Louis C.: quoted on water mills in North America, 31; quoted on water-power potential, 40
Hurricane Creek: 88
Hutchens, Issac: 110
Hutchens, John: 110
Hutchens, Wilson: 110
Hutcheson, Mary: 135
Hutcheson, Robert Samuel: 134
Huzzah Creek: 69, 70

Illinois: 39
Indian Territory: 106
Ireland: 18
Isbell, George: 106

Isbell, John: 104, 106
Isbell, Thomas: 104
Isbell Mill: *see* Jolly Mill
Isom, Rhuhama J.: 64

J. F. Smith Mills: 130
Jack's Fork River: 43–45
Jackson, Mo.: 50
Jefferson County, Mo.: 53, 59, 60, 146
Jesus, quoted on millstones: 6
Johnson, Linda: 84
Johnson Spring Mill: 88
Jollification, Mo., destruction of, in Civil War: 106
Jolly, Mo.: 106
Jolly Mill: 155, 164; construction of, 104; location of, 104, 106; restoration of, 107; changing ownership of, 106–107; present status of, 162
Jolly Mill Park Foundation: 162
Jolly Rolling Mill: 107
Joplin, Mo.: 39

Kansas: 39
Kentucky: 40
Kephart, Horace M., quoted on "going to mill": 149–50

Kickapoo Indian Trail: 128
Kimmel Mill: 118
King Ethelbert of Kent: 27
Klemme, Lester: 70; as entrepreneur, 146–47
Klemme's Dillard Roller Mill: 72
Klepsig, Charles: 44
Klienschmidt, Charles: 56

Laclede County, Mo.: 80, 111
Lalumondiere, James: 57, 62
Lebanon, Mo.: 80, 85, 130, 156
Leo A. Drey Foundation: 72
Lewis, Edna: 82
Lilly White (brand): 56, 57
Little North Fork River: 94
Longrun, Mo.: 96
Lorimier, Don Louis: 47
Lucas, A. C.: 107
Luther, Mo.: 115

McCaskill, George: 44
McDowell Mill: 108, 164; location of, 108, 110; unusual construction of, 110–11; decline of, 111

McElroy, John: 80
McLain, Alfred: 118
McLain Mill: *see* Old Appleton Mill
McLane, James W.: 119
Maddox, Thomas: 60
Maddox Mill: 60, 61
Mainprize, George W.: 92, 93
Manchester, David: 53
Marble Hill, Mo.: 6, 10, 12, 76
Mark Twain National Forest: 86, 89, 93
Matthew Ritchey Mill: *see* Ritchey Mill
Meyer, Theodore W.: 119, 120
"Millers' Bible": 37
Mills, Nellie Alice, quoted on excursion to mill: 160
Mills, early examples of: 15–16
Mill technology: contribution to, of Greeks, 16, 18; Roman contributions to, 22, 24; expanding nature of, 29–30; state of, in Medieval Europe, 30; state of, in colonial North America, 31–32; advancement of, in turbines, 34–35; advancement of, by automation, 37; movement of, into Missouri Ozarks, 38; impact on water mills, 144
Mischke, Emil: 69, 70
Mischke, Mary: 69

Mississippi River: 39, 47, 104, 148
Missouri: 14, 31, 32, 39, 47, 56, 147, 161
Missouri Department of Natural Resources: 51, 72, 162
Missouri Iron and Steel Company: 93
Missouri Ozarks: 13, 31, 32, 38, 46, 99, 103, 108, 144; as source of railroad ties, 52
Missouri River: 39
Monett, Mo.: 159
Montauk Mill: 131
Morrison, A. P.: 138; as entrepreneur, 140

National Geographic: 103
National Park Service: 46
National Register of Historic Places: 107, 115
Neosho River: 39
New England: 130
Newton County, Mo.: 104, 107, 123
Newtonia, Mo.: 106
Norford Lake: 63
Norse mill: description of, 18–19; spread of, 18–19
North America: 14, 30, 31, 38
North Carolina: 40, 47, 74, 90

North Fork River: 63, 68, 132
Norway: 18
Noser, Edward: 115
Noser, John J.: 114, 115
Noser Mill: 63, 138, 155, 161, 164; location of, 112; construction of, 112–14; changing use of, 115; present status of, 161

Oklahoma: 39, 84
Old Appleton Mill: 63, 118, 138, 153, 155, 164; early days of, 118; innovations to, 119; changing ownership of, 119; alterations in, 120; decline of, 120, 122; location of, 124; destruction of, by flood, 163
Old Mill Lodge: 147
Old Red Mill: *see* Alley Spring Mill
O'Neal, Joe W.: 136
Oregon County, Mo.: 74, 86, 90
Osage Fork River: 80, 156
Ozark County, Mo.: 96, 101–103, 137, 152
Ozark Plateau: 116
Ozark Queen Flour (brand): 96
Ozarks region: definition of, 39; described, 39–40

Parham, Allen: 80
Paris, France: 28
Pennsylvania: 31
Perry County, Mo.: 116, 118
Picturesque America (book): 3
Pioneer, Mo.: 123
Prussia: 76
Purdy, Mo.: 110
Pyrenees Mountains: 102

Radeacker, Albert: 62
Radeacker, Louis: 60
Radeacker, Walter: 62
Radeacker, Wilbert: 62
Radeacker, William: 62
Reynolds, Terry S.: 26; quoted on waterpower, 30; quoted on horizontal mill, 165–66n.
Ritchey, Matthew: 125
Ritchey, Mo.: 123, 160, 161
Ritchey Mill: 155, 158, 164; location of, 123; construction of, 125; destruction of second, 126; production in, 126–27; recreational use of, 127; present status of, 163–64

Ritchey Milling Company, incorporation of: 125–26
Roman Empire: expansion of, 24; collapse of, 27
Roman mill: description of, 16; spread of, 21, 24
Romans, innovations on Greek mill: 22

S. R. Burford and Company, founding of: 50
Saint Louis, Mo.: 93
Scandinavia: 28
Schlicht, Charles: 131
Schlicht, John: 130
Schlicht, Sherman: 131
Schlicht, William: 131
Schlicht Mill: 12, 145, 153, 164; location of, 128; origin of, 128, 130; improvements to, 130; as a social, commercial center, 130–31; changing ownership of, 131; neglect of, 163
Schulze, Arthur: 120
Scotland: 18
Sedgewickville, Mo.: 78
Seine River, France: 28

Shannon County: 43
Shoal Creek: 123, 125, 126, 155, 158, 160, 161
Shoemaker, George: 138
Silverman, L.: 50
Slovak, Emil: 89
Smith, Tay: 68
South Carolina: 40
Southerland, Bill: 136
Springfield, Mo.: 104, 126
Springfield, Ohio: 56
Squires, John: 94
Starnes, Bill: 82
Stout, Douglas: 82
Strain, Joseph: 128; as entrepreneur, 145–46
Swedeborg, Mo.: 128
Sweden: 32
Sycamore, Mo.: 66, 101, 137

Talley, John: 135
Taylor, Tell: 6
Tecumseh, Mo.: 63
Tennessee: 40, 90, 130
Thomas, G. E.: 127
Thomas, Gordon: 127, 158, 161
Topaz, Mo., rise and fall of: 132
Topaz Mill: 37, 40, 86, 148, 155, 156, 164; location of, 132; origin of, 132, 134; developments at, 134–35; decline of, 135, 136; present status of, 163–64
Troxel, Manford: 126; recollections of, 155
Tub mill, description of: 32, 34
Turbine, description of: 34–35

Union, Mo.: 112
Unterreiner, Leo F.: 119, 120

Vandivort, Paul: 50
Viburnum, Mo.: 69, 72
Viehland, Erwin: 62
Voire River, France: 27
Voss, Dietrich: 112

Wales: 27
Water mills: influence of, 3–8; aesthetic quality of, 3–4; effects on language, 5–6; effects on Jake K. Wells, 8–13;

historical origin of, 14–15; significance to North America, 14; precursors of, 15–16; basic idea of, 15–16; spread of, 27–28; automation of, 37; proliferation of, in Ozarks, 40; importance of, to Ozarks region, 108; and changing technology, 143–44; and entrepreneurs, 145–47; frontier use of, 149–50; estimates of, in Missouri, 1840–1880, 152; as a social center, 152–53; as centers of enterprise, 153, 154; and town development, 155–56, 158; importance of, to towns, 156, 158; as promoters of social life, 158–59, 161; *see also* Greek mill; Norse mill; Roman mill; and tub mill, description of

Waterpower: increasing application of, 29–30; developments in application of, 30

Waterpower: A History of Industrial Power in the United States, 1780–1930 (book): 31

Weaver, A. O.: 153; quoted on "going to mill," 150

Webster Groves, Mo.: 70

Wells, Jake K.: 44, 50, 52, 54, 60, 62, 64, 66, 70, 76, 78, 82, 88, 93, 96, 98, 101, 103, 107, 111, 112, 120, 126, 135, 164; influence of grandfather on, 8; influence of water mills on, 8, 10; quoted on artistic interests, 10; poetic reaction to water mills, 12–13

White River: 63, 94, 132

Whitewater River: 47, 49–52, 74, 116

William the Conqueror: 28

Williams, S. J.: 94

Wisdom, Francis: 69

Wisdom Mill, destruction of: 69

World War I: 107

World War II: 57, 125, 140

Yerkes Mill: 54

Young Mill-Wright and Millers' Guide (book), importance of: 37–38

Zanoni Mill: 10, 140, 152, 155, 156, 164; location of, 102, 137; description of, 137–38; origin of, 138–40; decline of, 140–41; as example of old water mills, 144

Zanoni Spring, capacity of: 137